NORWICH CITY COLLEGE LIBRARY		
Stock No.	185455	
Class	624.1771 MAC	
Cat.		Proc.

D1579499

Design of Structural Masonry

W.M.C. _____ B.Sc., Ph.D., C.Phys., M_____st.P., C.Eng.
_____er *University, Edinburg_*

palgrave

185 455

© W. M. C. McKenzie 2001

All rights reserved. No reproduction, copy or transmission of
this publication may be made without written permission.

No paragraph of this publication may be reproduced, copied or
transmitted save with written permission or in accordance with
the provisions of the Copyright, Designs and Patents Act 1988,
or under the terms of any licence permitting limited copying
issued by the Copyright Licensing Agency, 90 Tottenham Court Road,
London W1T 4LP.

Any person who does any unauthorised act in relations to this
publication may be liable to criminal prosecution and civil
claims for damages.

The author has asserted his rights to be identified as the author of this
work in accordance with the Copyright, Designs and Patents Act 1988.

First published 2001 by
PALGRAVE
Houndmills, Basingstoke, Hampshire RG21 6XS and
175 Fifth Avenue, New York, N.Y. 10010
Companies and representatives throughout the world.

PALGRAVE is the new global academic imprint of St. Martin's Press LLC
Scholarly and Reference division and Palgrave Publishers Ltd (formerly
Macmillan Press Ltd).

ISBN 0–333–79237–8

This book is printed on paper suitable for recycling and
made from fully managed and sustained forest sources.

A catalogue record for this book is available
from the British Library.

10	9	8	7	6	5	4	3	2	1
10	09	08	07	06	05	04	03	02	01

Printed in Great Britain by
Antony Rowe Ltd, Chippenham,
Wiltshire.

Contents

Preface

Rationale

Existing design textbooks for undergraduate engineering students neglect, to a large extent, the importance of masonry as a structural building material. As a consequence, relatively few textbooks provide information on the design of masonry structures. *Design of Structural Masonry* has therefore been written to:

♦ provide a comprehensive source of information on practical masonry design,
♦ introduce the nature and inherent characteristics of masonry given in relation to the requirements of BS 5628,
♦ introduce the use of Eurocode EC6 in structural masonry design.

The book's content ranges from an introduction to masonry as a material to the design of realistic structures including and beyond that usually considered essential for undergraduate study.

Readership

Design of Structural Masonry is written primarily for undergraduate civil and structural engineers. The book will also provide an invaluable reference source for practising engineers in many building, civil and architectural design offices.

Worked Examples and Review Problems

The design of structures/elements is explained and illustrated using numerous detailed, relevant and practical worked examples. These design examples are presented in a format typical of that used in design office practice in order to encourage students to adopt a methodical and rational approach when preparing structural calculations.

Review problems are included at the end of each chapter to allow the reader to test his/her understanding of the material; references are given to the relevant sections in the preceding chapter.

Design Codes

It is essential when undertaking structural design to make frequent reference to the appropriate design codes. Students are encouraged to do this whilst using this text. It is assumed that readers will have access to either *Extracts from British Standards for Students of Structural Design* (which is a standard text in virtually all undergraduate structural design courses) or the complete versions of the necessary codes, with the exception of the Eurocode EC6. In Chapter 6 where EC6 is discussed, appropriate extracts from the code are given to illustrate where variables, etc., have been obtained.

W.M.C. McKenzie

To David, Myra and Maureen

Acknowledgements

I wish to thank Christopher Glennie of Palgrave and Tessa Handford for their help and advice during the preparation of this text. I am indebted to Mr. Michael Hammett and 'The Brick Development Association', and Mr. Colin Baxter, for the use of the cover photographs illustrating Winterton House and Edinburgh Castle respectively. Finally, I wish to thank Caroline once again for her endless support, encouragement and proof-reading.

1. Structural Masonry

Objectives: *'To introduce masonry, i.e. brickwork, blockwork and natural stone as a structural material and discuss relevant material characteristics required for limit-state design.'*

1.1   Introduction

The use of masonry for construction during many centuries includes the building of the *pyramids* at Giza in Egypt, the *Great Wall* of China, the *temples* and *palaces* of the Incas in Peru and numerous *baths, amphitheatres* and *aqueducts* of the Roman Empire. The bible refers to bricks and mortar in Genesis Chapter 1, Verse 3: *"And they said to one another 'Come, let us make brick and burn them thoroughly'.......and they had brick for stone, and slime had they for mortar."* Design techniques based on well-established scientific principles, however, have only been developed during the latter part of the 20th. century.

The earliest forms of masonry were made from mud and straw hand-moulded bricks, which were dried in the sun. The manufacture of these *'structural units'* did not require the use of tools and they were used from about 6000 BC.

During the Bronze Age, about 3000 BC, the development of purpose-made tools enabled stone, previously used in its natural form, to be cut and shaped more readily for construction purposes.

In Mesopotamia around 2500 BC, the techniques used for making kiln-fired pottery were adopted and applied to brick making. This resulted in durable fired-clay bricks which had been manufactured in purpose-made moulds. Clay, a sedimentary deposit, is a hydrated silicate of alumina mixed with various impurities. The clay used for the manufacture of bricks is fine-grained and sufficiently plastic when wet that it can be moulded. During the firing process the chemical and physical structure of clay is altered resulting in a hard, coherent mass.

The art of brick making continued throughout history until the fall of the Roman Empire, after which it was lost for a period of time. A revival occurred in the 13th. century and continued until the present day. During that period many major structures have been constructed in masonry, e.g. Hampton Court in England in the early 16th. century and numerous cathedrals throughout the UK and Europe. The use of timber in house construction in the UK was replaced by brickwork following the Great Fire of London in 1666.

The mechanisation and development of brick making occurred in the mid-19th. century. Prior to this time the firing of bricks had always been in intermittent kilns. Using this technique, moulded and partially dried bricks were loaded into a kiln and fired. On completion of the firing the fire was put out, the kiln opened and the bricks allowed to cool. This process was then repeated for the next batch.

Modern brick making is carried out using a continuous process in which batches of bricks are loaded, fired, cooled and removed in permanent rotation. The shaping of clay to produce bricks is carried out either by extrusion or by moulding/pressing:

♦   *Extrusion*: This technique is used with clays, which develop high levels of
    plasticity. The clay is extruded into continuous columns, which are
    subsequently wire-cut into individual brick lengths as shown in
    Figure 1.1. The individual bricks are dried before firing.

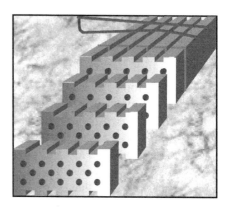

clay extruded into columns   columns wire-cut into bricks

Figure 1.1

♦   *Moulding/Pressing*: This technique is used for clays, which only develop low
    levels of plasticity. The clay is either placed directly into a
    mould whilst still comparatively dry, or is formed into 'clots'
    with a dough-like consistency and dropped into a mould and
    pressed as shown in Figure 1.2.

clay pressed into moulds   bricks ready for firing

Figure 1.2

The strength of masonry/brickwork is dependent on a number of factors one of which is the unit strength. (**Note**: the distinction between *brickwork* – an assemblage of bricks and mortar and *brick* – the individual structural unit. In this text in general, reference to bricks and brickwork also implies blockwork, stonework etc.).

In civil engineering projects which require high strength characteristics, high density *engineering bricks* are frequently used whilst in general construction *common bricks* (commons) are used. Where appearance is a prime consideration *facing bricks* are used combining attractive appearance, colour and good resistance to exposure. Bricks, which are non-standard size and/or shape, are increasingly being used by architects and are known as *specials* (see Figure 1.3*)*.

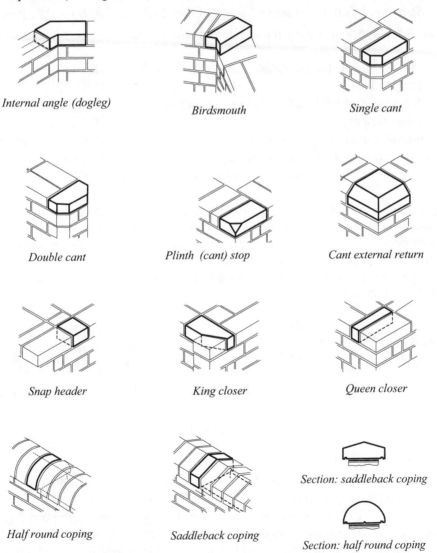

Internal angle (dogleg)   Birdsmouth   Single cant

Double cant   Plinth (cant) stop   Cant external return

Snap header   King closer   Queen closer

Half round coping   Saddleback coping   Section: saddleback coping

Section: half round coping

Figure 1.3 A selection of *special* bricks

The design of structural masonry/brickwork in the U.K. is governed by the requirements of BS 5628 - Code of Practice for the Use of Masonry Parts 1, 2 and 3.

The references to Clause numbers in this text relate to BS 5628 : Part 1 : 1992 unless stated otherwise.

1.2   Materials

Masonry can be regarded as an assemblage of structural units, which are bonded together in a particular pattern by mortar or grout. Masonry may be unreinforced, reinforced or pre-stressed; each of these is discussed in detail in Chapters 2, 3, 4 and 5.

1.2.1   *Structural Units: (Clause 7 and Section 2 Clause 5 of BS 5628 : Part 3)*

There are seven types of structural unit referred to in BS 5628 they are:

- ♦ calcium silicate (sandlime and flintlime) bricks   (BS 187),
- ♦ clay bricks   (BS 3921),
- ♦ dimensions of bricks of special shapes and sizes   (BS 4729),
- ♦ stone masonry   (BS 5340),
- ♦ precast concrete masonry units   (BS 6073 : Part 1),
- ♦ reconstructed stone masonry units   (BS 6457),
- ♦ clay and calcium silicate modular bricks   (BS 6649).

The specification for each of these unit types is given in the appropriate British Standard as indicated. The selection of a particular type of unit for any given structure is dependent on a number of criteria, e.g. strength, durability, adhesion, fire resistance, thermal properties, acoustic properties and aesthetics.

The structural units may be solid, solid with frogs, perforated, hollow or cellular as indicated in Figure 1.4.

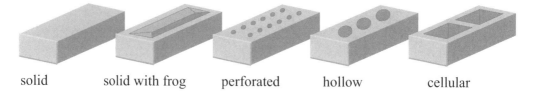

solid   solid with frog   perforated   hollow   cellular

Figure 1.4   Types of structural unit

1.2.1.1 Dimensions and Sizes

The specifications for the sizes of clay bricks, calcium silicate (sandlime and flintlime) bricks and precast concrete masonry units are given in BS 3921 : 1985, BS 187 : 1978 and BS 6073 : Part 1 : 1981 respectively. The sizes of bricks are normally referred to in terms of work sizes and co-ordinating sizes as shown in Figure 1.5. When using clay or calcium bricks the standard work sizes for individual units are 215 mm length × 102.5 mm width × 65 mm height. In most cases the recommended joint width is 10 mm resulting in co-ordination sizes of bricks equal to 225 mm × 112.5 mm × 75 mm.

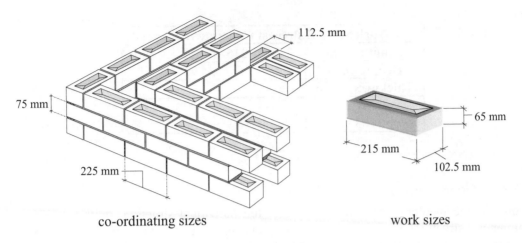

co-ordinating sizes   work sizes

Figure 1.5

When designing it is more efficient and economic to specify dimensions of masonry minimising the cutting of brickwork. Wherever possible the dimensions of openings, panels returns, piers etc. should be a multiple of the co-ordinating size, plus or minus the joint thickness where appropriate. Brickwork dimension tables are available from the Brick Development Association (1).

In BS 6073 : Part 1 : 1981, '*work sizes*' are given for precast concrete masonry blocks and bricks and are shown in Figures 1.6 and 1.7.

Work Sizes of Precast Concrete Blocks								
Length - mm	390	440	440	440	440	590	590	590
Height - mm	190	140	190	215	290	140	190	215
Thickness - mm								
60	*	*	*	*	*			
75	*	*	*	*	*	*	*	*
90	*	*	*	*	*	*	*	*
100	*	*	*	*	*	*	*	*
115	*			*				
125				*				*
140	*	*	*	*	*	*	*	*
150	*	*	*	*	*	*	*	*
175				*				*
190	*	*	*	*	*	*	*	
200	*	*		*	*	*	*	*
215			*	*	*	*	*	*
220			*	*				
225		*		*				*
250				*				*

Figure 1.6

Work Sizes of Precast Concrete Bricks				
Length - mm	290	215	190	190
Height - mm	90	65	90	65
Thickness - mm				
90	*		*	*
103		*		

Figure 1.7

1.2.2   *Mortar: (Clause 15)*

Mortar is the medium which binds together the individual structural units to create a continuous structural form e.g. brickwork, stonework etc. Mortar serves a number of functions in masonry construction, i.e. to:

- bind together the individual units,
- distribute the pressures evenly throughout the individual units,
- infill the joints between the units and hence increase the resistance to moisture penetration,
- maintain the sound characteristics of a wall,
- maintain the thermal characteristics of a wall.

Early mortars which bonded together kiln-burned bricks included bitumen (e.g. in the Ziggurat at Ur in Iraq), and later from the 1st. century onwards the lime-based mortars developed by the Romans. The mortars used by the Romans contained water, lime, sand and pozzolana (hydraulic cement), and proved to be very durable and effective. Similar mortars used by builders in Northern Europe during and after the mediaeval periods were found to be less durable. This was largely due to the lack of control of both the burning of the lime before use and the proportions used; unlike the Romans who scientifically controlled the details of the materials used, the Northern Europeans were less particular.

Present day mortars are specifically manufactured to suit the type of construction involved. In most cases they are mixtures of sand, cement and water. The workability is often improved by the inclusion of lime or a mortar plasticiser. Lime is used in mortar for several reasons:

- to create a consistency which enables the mortar to *'cling and spread'*,
- to help retain the moisture and prevent the mortar from setting too quickly,
- to improve the ability of the mortar to accommodate local movement.

Modern mortars containing lime should not be confused with lime mortars. True lime mortars are mortars in which lime is used instead of Portland cement as the primary binder material. There are two types, hydrated (non-hydraulic) lime and hydraulic lime mortars. The set and strength characteristics of each type are different. The physical properties of lime mortars have not been quantified as comprehensively as those of widely used

Portland cement mortars and they should only be specified after careful consideration of their intended use and suitability. They should not be used as direct substitutes for Portland cement mortars.

Plasticisers can be used with mortars which have a low cement : sand ratio to improve the workability. Their use introduces air bubbles into the mixture which fill the voids in the sand and increase the volume of the binder paste. The introduction of plasticisers into a mix must be carefully controlled since the short-term gain in improved workability can be offset in the longer term by creating an excessively porous mortar resulting in reduced durability, strength and bond. This is emphasised in BS 5628 : Part 1 : 1992, Clause 17 in which it is stated: *'Plasticisers can only be used with the written permission of the designer.'*

The use of colouring pigments to produce coloured mortars is permitted as indicated in Clause 16 of BS 5628 : Part 1 : 1992: *'Pigments should comply with the requirements of BS 1014 and should not exceed 10% by mass of the cement in the mortar. Care should be taken to ensure that the pigment is evenly distributed throughout the mortar and that the strength of the mortar remains adequate. Carbon black should be limited to 3% by mass of the cement.'* (**Note:** BS 1014 'Specification for pigments for Portland cement products').

To date there are no known admixtures, which effectively provide frost protection for mortars without introducing other undesirable effects. This is particularly the case with admixtures based on calcium chloride. In Clause 18 of the code, the use of this frost inhibitor is specifically prohibited: *'The use of calcium chloride or frost inhibitors based on calcium chloride is not permitted in mortars.'*

The use of ready-mixed mortars is permitted but is controlled as indicated in Clause 15.2 of the code: *'Ready-mixed lime : sand for mortar should comply with the requirements of BS 4721. The appropriate addition of cement should be gauged on site. Wet ready-mixed retarded cement : lime : sand mixes should be used only with the written permission of the designer.'* (**Note:** BS 4721 'Ready-mixed lime : sand for mortar').

The requirements for mortars in relation to strength, resistance to frost attack during construction and improvement in bond and consequent resistance to rain penetration are given in Table 1 of BS 5628 : Part 1. Four mortar designation types (i), (ii), (iii) and (iv) are specified in terms of their cement, lime, sand and plasticiser content and appropriate 28-day strengths are given. The mortar type is subsequently used in design calculations to determine characteristic masonry strengths (f_k).

1.2.3   Bonds:(Appendix B of BS 5628 : Part 3 : 1985)

Walling made from regular shaped units is constructed by laying the units in definite, specific patterns called bonds, according to the orientation of the long sides (*stretchers*) or the short sides (*headers*).

The method of laying structural units is specified in Section 8 of BS 5628 : Part 1 and detailed in Section 32 of Part 3 of the code. Normally all bricks, solid and cellular blocks are laid on a full bed of mortar with all cross joints and collar joints filled. (*A cross joint is a joint other than a bed joint, at right angles to the face of a wall. A collar joint is a continuous vertical joint parallel to the face of a wall.*)

In situations where units are laid on either their stretcher face or end face, the strength of the units used in design should be based on tests carried out in this orientation. In bricks with frogs the unit should be laid with the frog or larger frog uppermost. The position and filling of frogs is important since both can affect the resulting strength and sound insulation of a wall. Cellular bricks should be laid with their cavities downwards and unfilled.

It is essential when constructing brickwork walls to ensure that the individual units are bonding together in a manner which will distribute the applied loading throughout the brickwork. This is normally achieved by laying units such that they overlap others in the course below. The resulting pattern of brickwork enables applied loads to be distributed both in the horizontal and vertical directions as shown in Figure 1.8(a) and (b).

(a) Distribution due to in-plane loading   (b) Distribution due to out-of-plane loading

Figure 1.8

A number of bonds have been established which provide brickwork walls with the required characteristics, i.e.:

   ♦ vertical and horizontal load distribution for in-plane forces,
   ♦ lateral stability to resist out-of-plane forces,
   ♦ aesthetically acceptable finishes.

In BS 5628 : Part 3 masonry bonds are defined for both brickwork and blockwork.
The seven bonds indicated for **brickwork** are as shown in Figures 1.9 to 1.15.

   ♦ *English bond:*   shows on both faces alternate courses of headers and stretchers,

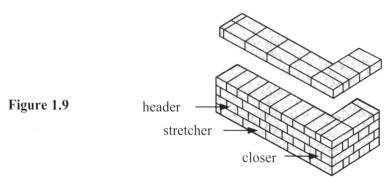

Figure 1.9   header

                      stretcher

                          closer

♦ *Flemish bond:*   shows on the face alternate headers and stretchers in each course. It may be built as a 'single Flemish bond', which shows Flemish bond on both faces of the wall,

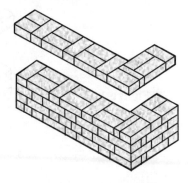

Figure 1.10

♦ *English garden-wall bond:*   shows a sequence of three courses of stretchers laid with half-lap to one course of headers,

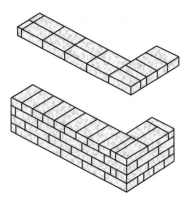

Figure 1.11

♦ *Flemish garden-wall bond:*   shows on both faces a sequence of three stretchers
(Sussex garden-wall bond)   to one header in each course of a full brick wall. In thicker walls, one face is formed in English bond,

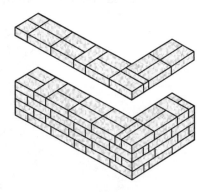

Figure 1.12

♦ *Heading bond:*   consists of bricks with their ends showing on the face of the
  (header bond)   wall, laid with a half lap of the brick width,

Figure 1.13

♦ *Quetta bond:*   this is used for walls a minimum of one and a half bricks thick. It
                   consists of alternate stretchers and headers arranged to leave a
                   series of vertical voids in the wall thickness. Vertical
                   reinforcement is placed in the voids, which are filled with grout
                   or fine concrete as the work proceeds,

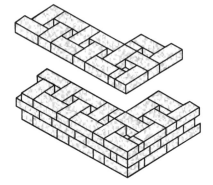

Figure 1.14

♦ *Rat-trap bond:*   shows bricks laid on edge, each course consisting of alternate
                     headers and stretchers. It has a similar appearance to Flemish
                     bond and may be vertically reinforced in the same way as Quetta
                     bond,

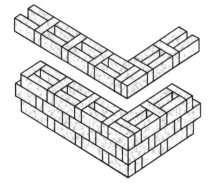

Figure 1.15

The three bonds for **blockwork** are as shown in Figures 1.16 to 1.18.

♦ *Running (stretcher) bond:*   this requires the block thickness to be equal to half of the block length. There are half blocks at the wall ends,

Figure 1.16

half blocks

♦ *Thin stretcher bond:*   this requires cut-block closers or quoins at the corner and half blocks at wall ends,

Figure 1.17

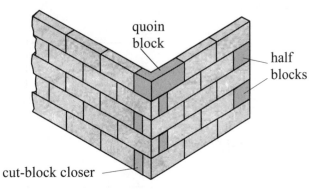

quoin block

half blocks

cut-block closer

♦ *Off-centre running bond:*   this requires three-quarters or two-thirds cut blocks at wall ends,

Figure 1.18

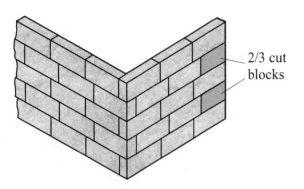

2/3 cut blocks

1.2.4   Joint Finishes: (Appendix B of BS 5628 : Part 3 : 1985)

The final appearance of brickwork is dependent on the finish of the joints between individual units (perpend/vertical-cross joints), and the bed joints between the courses. Various joint finishes can be created depending upon the desired aesthetic effect. If the finishes are created during construction then the process is called '*jointing*'; if this is done after completion of the brickwork it is called '*pointing*'. As indicated in Clause 27.7 of BS 5628 : Part 3, jointing is preferable since it leaves the bedding mortar undisturbed.

A number of joint finishes, as shown in Figures 1.19 to 1.22, are illustrated in Appendix B of BS 5628 : Part 3.

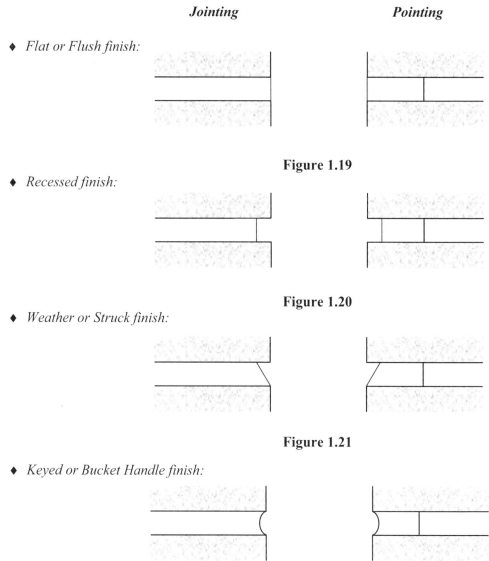

Jointing   *Pointing*

♦ *Flat or Flush finish:*

Figure 1.19

♦ *Recessed finish:*

Figure 1.20

♦ *Weather or Struck finish:*

Figure 1.21

♦ *Keyed or Bucket Handle finish:*

Figure 1.22

The type of joint finish selected will be influenced by a number of factors such as exposure conditions and aesthetics. The most effective types to resist rain penetration are weather-struck and keyed finishes. In situations where wind-driven rain is likely, finishes which produce a water-retaining edge, e.g. recessed, should be avoided.

The use of pointing is more prevalent when refurbishing existing brickwork. The mortar joints are normally raked out to a depth of approximately 15 mm and the brickwork brushed and wetted before refilling with the desired mortar. If raking out of a joint is carried out as a finishing feature, the reduction in wall thickness should be considered when calculating design stresses as indicated in a footnote to Clauses 8.2, 8.3 and 8.4 of BS 5628 : Part 1 : 1992.

1.2.5   *Damp-Proof-Courses:(Clause 12 and Clause 10 of BS 5628 : Part 3 : 1985)*

The purpose of a damp-proof course (d.p.c.) is to provide an impermeable barrier to the movement of moisture into a building. The passage of water may be horizontal, upward or downward. The material properties required of d.p.cs are set out in BS 5628 : Part 3, Clause 21.4.3, they are:

♦ an expected life at least equal to that of the building,
♦ resistance to compression without extrusion,
♦ resistance to sliding where necessary,
♦ adhesion to units and mortar where necessary,
♦ resistance to accidental damage during installation and subsequent building operations,
♦ workability at temperatures normally encountered during building operations, with particular regard to ease of forming and sealing joints, fabricating junctions, steps and stop ends, and ability to retain shape.

There is a wide range of materials which are currently used as d.p.c.s, they fall into three categories:

♦ flexible   e.g.   lead, copper, polyethylene, bitumen, bitumen polymers,
♦ semi-rigid e.g.   mastic asphalt,
♦ rigid   e.g.   epoxy resin/sand, three courses of engineering brick, and bonded slate.

The physical properties and performance of each type are given in BS 5628 : Part 3, Table 12. The relevant British Standards, which apply to each type are:

♦ bitumen   (BS 6398),
♦ brick   (BS 3921),
♦ polyethylene   (BS 6515),
♦ all others   (BS 743).

The correct positioning of d.p.cs is important to ensure continuity of the impervious barrier throughout a building. In Clause 21.5 of BS 5628 : Part 3, detailed advice is given relating to the use of d.p.c.s in most common situations, i.e. below ground level, immediately above ground level, under cills, at jambs of openings, over openings, at balcony thresholds, in parapets and in chimneys. A typical detail for a steel lintol over an opening in a cavity wall is shown in Figure 1.23.

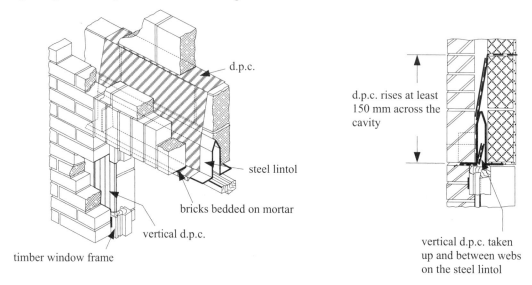

Figure 1.23

1.2.6   *Rendering:(Clause 21.3.2 of BS 5628 : Part 3 : 1985)*

The rendering of masonry involves applying an additional surface to the external walls to improve the weather resistance and in some instances to provide a decorative finish. There are five commonly used types of render, they are:

◆ pebbledash (dry dash)	this produces a rough finish with exposed stones which are thrown onto a freshly applied coat of mortar,
◆ roughcast (wet dash)	this produces a rough finish by throwing a wet mix containing a proportion of small stones,
◆ scraped or textured	the final coat of mortar is treated using one of a variety of tools to produce a desired finish,
◆ plain coat	the final coat is a smooth/level coat finished with wood-cork or felt-faced pad,
◆ machine applied finish	the final coat is thrown on the wall by a machine producing the desired texture.

It is important when applying a rendered finish, to give consideration to the possibility of cracking. In circumstances where strong, dense mixes are used for the render to protect against severe exposure conditions, there is the risk of considerable shrinkage cracking. The tendency is to use weaker renders based on cement : lime mixes which accommodate higher levels of shrinkage, are more absorbent and reduce the flow of water over the surface cracks.

1.2.7 Wall Ties:(Clause 13 and Clause 19.5 of BS 5628 : Part 3 : 1985)

Wall ties are required to tie together a masonry wall and another structural component/element such that they behave compositely to resist/transfer the applied loading. The design of wall ties should comply with the requirements of **'BS 1243 Specification for metal ties for cavity wall construction'**. In situations where corrosion is likely, the ties should be manufactured from either stainless steel or a non-ferrous material. There are a number of proprietary types of wall tie which are available for various purposes some of which are indicated in Figures 1.24 to 1.28.

♦ Brick-to-brick

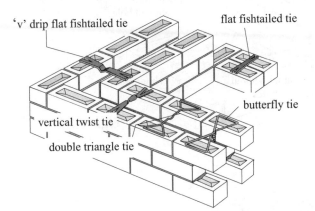

'v' drip flat fishtailed tie
flat fishtailed tie
butterfly tie
vertical twist tie
double triangle tie

Double triangle and butterfly ties, which are more flexible than the others, are more suited to domestic scale construction.

Vertical twist and fishtailed ties are more substantial and are generally more suitable for highly stressed masonry.

Figure 1.24

Some of these ties must have a 'drip' to prevent the movement of water through a cavity (as shown in Figure 1.25) and be designed such that they minimise the retention of 'mortar droppings' during construction.

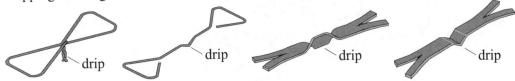

drip drip drip drip

Figure 1.25

♦ Brick/block-to-timber

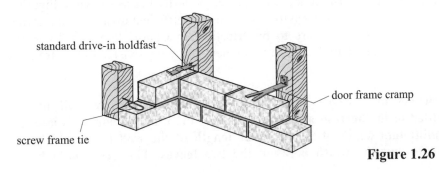

standard drive-in holdfast
door frame cramp
screw frame tie

Figure 1.26

♦ Brick-to-block/concrete

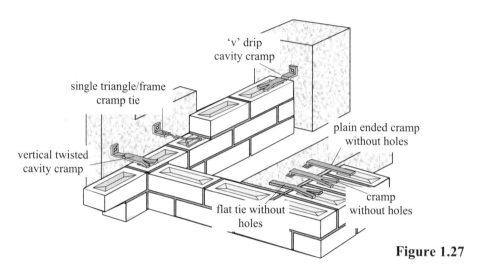

Figure 1.27

♦ Brick/block-to-steelwork

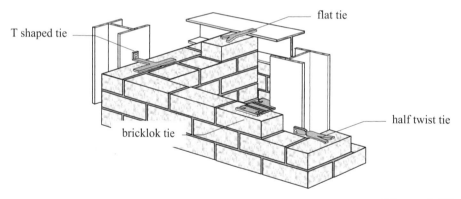

Figure 1.28

The most commonly used wall ties are those used in cavity-wall construction (see section 1.3.4) i.e. vertical twist, double triangle and butterfly ties as shown in Figure 1.24. The strength and spacing of ties in cavity walls must be sufficient to develop the combined stiffness of both leaves if they are to be effective. In Clause 19.5 and Table 9 (see Figure 1.29) of BS 5628 : Part 3 : 1985 requirements for the provision of ties are given as indicated:

'The leaves of a cavity wall should be tied together by wall ties imbedded in the horizontal mortar joints at the time the course is laid, to a minimum depth of 50 mm. The length of the wall tie should be chosen to suit the width between the two leaves. The ties should be

placed at a frequency of not less than the values given in table 9(A) and they should be staggered and evenly distributed. Additional ties should be provided within 225 mm of all openings so that there is one for each 300 mm of height of the opening. Consideration should be given to providing additional flexible ties across the cavity adjacent to movement joints.

The choice of the type of tie depends on the cavity width see table 9(B). In situation of severe or very severe exposure as defined in 21.2*, copper alloy or stainless steel ties should be used.'

Note: The exposure of walls to local wind-driven rain is classified in Table 10 of BS 5628 : Part 3 : 1985 varying from Very Severe to Very Sheltered and is dependent on the determination of a *'local wind-driven rain index'*; the evaluation of this is not covered in this text.

Extract from BS : 5628 : Part 3 : 1985

Table 9. Wall ties					
(A) Spacing of ties					
Least leaf thickness (one or both)	Type of tie	Cavity width	Equivalent no. of ties per square metre	Spacing of ties	
				Horizontally	Vertically
mm 65 to 90 90 or more	All See table 9 (b)	mm 50 to 75 50 to 150	4.9 2.5	mm 450 900	mm 450 450
(B) Selection of ties					
			Type of tie in BS 1243	Cavity width	
↑ Increasing strength	Increasing flexibility and sound insulation ↓		Vertical twist	mm 150 or less	
			Double triangle	75 or less	
			Butterfly	75 or less	

Figure 1.29

1.3   Structural Forms

Masonry is constructed by building up structural units in horizontal layers called *courses*, which are bonded together by intermediate layers of mortar. Walling made from stone is generally one of two types, rubble walls or ashlar walls, whilst walling from bricks/blocks can be one of many forms such as; solid, cavity, collar jointed, diaphragm, fin, reinforced or pre-stressed walls.

1.3.1   Rubble Walls

Rubble walls are built from irregular and coarsely jointed quarried stone. Overlapping stones in successive adjacent courses achieve longitudinal bond in such walls. The variation in size of stones used results in laps of differing, random lengths. Transverse bonding through the thickness of walls is achieved by using larger stones known as bonders. These are normally placed one for every square metre of wall in each face. To avoid the passage of moisture through the full width of a wall, bonders do not extend through the full width. The spaces between the lapping stones and bonders in each face are filled with small pieces of stone as shown in Figure 1.30. If courses are roughly levelled at regular intervals with more care being taken in positioning and bedding of stones a stronger wall will be produced. When laid without mortar such walling is known as dry masonry.

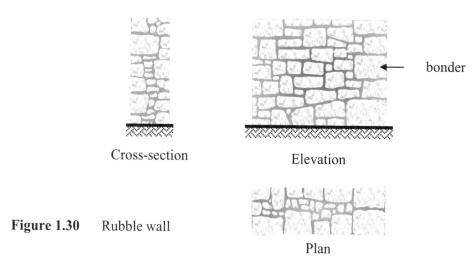

Cross-section   Elevation

Figure 1.30   Rubble wall

Plan

1.3.2   *Ashlar Walls*

Built from carefully shaped and set stone blocks with close, fine joints Ashlar walls are normally constructed with a brick backing using stone as a front face as shown in Figure 1.31. The stone blocks are cut to sizes which correspond to a set number of brickwork courses. It is important to protect the ashlar wall from any salts, which may leech from the brickwork and consequently any ashlar stones which are in contact with the backing wall should be painted with a coating of bitumen or proprietary alternative.

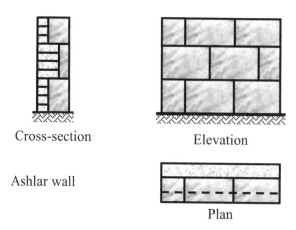

Cross-section   Elevation

Figure 1.31   Ashlar wall

Plan

1.3.3   Solid Walls and Columns

The form and construction of solid walls and columns can take one of numerous alternatives. Bonded single-leaf walls as shown in Figure 1.32 (a) are primarily used externally for earth retaining structures or boundary walls and internally for loadbearing walls supporting beams and/or floor and roof slabs. They can be of any thickness but typically are one half brick (102.5 mm), one brick (225 mm) or one-and-one half brick (327.5 mm) thick.

In situations where structural requirements dictate that a wall thickness should be greater than one half brick thick and architectural requirements dictate that fairfaced, stretcher bond brickwork is required on both faces, double-leaf (or collar jointed) walls as shown in Figure 1.32(b) are often used. The vertical joint between the two leafs is known as a '*collar joint*' and various conditions are specified in BS 5628 : Part 1 which must be satisfied if the wall is to be designed as a single-leaf, solid wall.

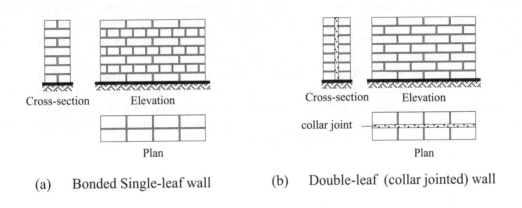

    (a)   Bonded Single-leaf wall   (b)   Double-leaf (collar jointed) wall

Figure 1.32

Faced walls as shown in Figure 1.33(a) comprise two different types of structural unit which are bonded together to form a solid wall. Usually they are used for architectural reasons where facing brick is required on one face only. Since facing bricks are more expensive than common bricks it is more economic to bond the outside layer into less expensive units. Care must be taken when designing such walls to ensure that the physical characteristics (e.g. thermal expansion/contraction, shrinkage properties etc.) of each type of unit are similar. The strength of the wall is based on the full width and on the weaker of the units used.

Unlike facing bricks, which are bonded into and contribute to the strength of a wall, a veneer facing applied to a wall for architectural reasons as shown in Figure 1.33(b), is not loadbearing. The veneer is tied to the loadbearing brickwork (i.e. not bonded) and composite action does not occur. In structural terms the veneer merely increases the dead load to be supported. As in the case of faced walls care must be taken to ensure compatability between the materials used.

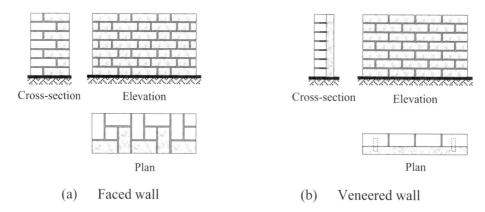

Cross-section   Elevation   Cross-section   Elevation

Plan   Plan

(a)   Faced wall   (b)   Veneered wall

Figure 1.33

Stiffened walls (see Figure 1.34(a)) are frequently used where concentrated loads (e.g. end reactions from beams/trusses etc.) are applied at intervals along the length of a wall and/or where a wall panel is required to resist lateral loading. Piers are bonded into the wall effectively increasing the stiffness and load carrying capacity.

Isolated vertical loadbearing members in which the width is not more than four times the thickness are referred to as columns. (**Note:** piers are members which form an integral part of a wall, in the form of a thickened section placed at intervals along the wall.) The majority of masonry columns are either square or rectangular in cross-section, however, other shapes can be adopted as shown in Figure 1.34(b).

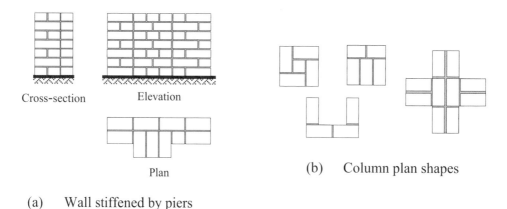

Cross-section   Elevation

Plan

(a)   Wall stiffened by piers

(b)   Column plan shapes

Figure 1.34

1.3.4   *Cavity Walls*

Cavity walls have been adopted universally as exterior walls in buildings to resist both vertical and lateral loading. They provide strength and stability, resistance to rain penetration, thermal and sound insulation in addition to fire resistance. The construction of a cavity wall normally comprises two parallel single-leaf walls at least 50 mm apart and tied together with metal ties (see section 1.2.7). The outer leaf is usually one half-brick thick with a similar inner leaf or an inner leaf of light-weight concrete blocks as shown in Figure 1.35(a) and (b).

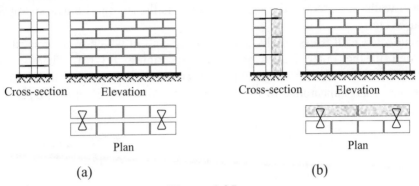

(a)　　　　　　　　　　　　　　(b)

Figure 1.35

The effect of the ties is to increase the stiffness of each individual leaf and consequently reduce its tendency to buckle under load. The width of the cavity between the leaves may vary from 50 mm to 150 mm but should not be greater than 75 mm if the thickness of one of the leaves is less than 90 mm. The minimum width is required to avoid inadvertent bridging (e.g. by mortar droppings) and the maximum width is to limit the free length of the ties and hence reduce their tendency to buckle when subjected to compressive forces.

The thermal insulation of cavity walls is frequently increased by the addition of non-loadbearing materials inserted in the cavity either during or after construction as shown in Figure 1.36(a).

The loadbearing capacity of cavity walls can be increased by incorporating piers in one of the leaves, see Figure 1.36(b), or by filling the cavity with concrete of 28-day strength not less than that of the mortar. In the latter case the wall can be designed as a solid, single-leaf wall of effective thickness equal to the overall actual thickness.

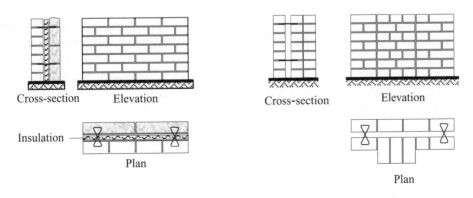

(a)　Cavity wall with insulation　　　(b)　Cavity wall with stiffened leaf

Figure 1.36

1.3.5   *Diaphragm and Fin Walls*

When it is desirable to construct a structural envelope around large, column-free spaces, diaphragm and fin walls provide an elegant solution. This is particularly so where tall perimeter walls are required, e.g. in sports halls, gymnasiums, swimming pools, warehouses etc.

A ***diaphragm wall*** is essentially a wide cavity wall in which transverse ribs have been bonded/tied to the outer leaves as shown in Figure 1.37(a). There are numerous geometrical arrangements possible within a cross-section which can be formulated to enhance the visual appearance of the wall and in addition accommodate service ducts as indicated in Figure 1.37(b).

(a)

(b)

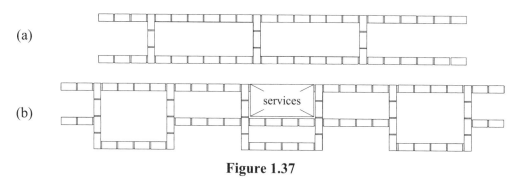

Figure 1.37

The cellular form of construction results in a series of ▮ or I sections in which the two parallel leaves behave as flanges resisting bending stresses and the transverse ribs behave as webs resisting shear forces (see Figure 1.38).

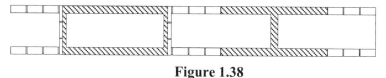

Figure 1.38

Clearly the connection between the ribs and the leaves is fundamental to the structural behaviour and adequate shear capacity must be provided either by using bonded brickwork or specially designed ties.

A ***fin wall*** is a stiffened wall in which the piers are extended as shown in Figure 1.39. It is a development of the diaphragm wall resulting in a series of '**T**' sections which resist the bending and shear stresses.

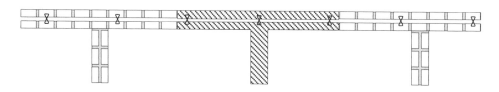

Figure 1.39

As with diaphragm walls careful consideration must be given to the shear stresses between the fin and the leaf to which it is bonded/tied. In both cases design is carried out on the assumption that the roof structure provides a propping force to the top of the wall. It is important to ensure that the roofing system adopted and its connection to the top of the wall are both capable of transferring the propping force to the appropriate vertical shear walls.

An added advantage of both systems over the more traditional steelwork/concrete frame structures is the low foundation pressures which enable nominal strip footings to be used.

1.3.6   *Reinforced and Prestressed Brickwork*

The inherent weakness of brickwork in resisting tensile stresses imposes restrictions on the efficiency which could otherwise be obtained when using masonry construction. As with concrete this problem can be overcome by introducing material, i.e. steel bars, to resist flexural tension and hence creating a reinforced section or alternatively, by introducing a pre-stress to eliminate the tensile stresses induced by loading.

Despite being introduced during the 19th. century, the use of steel to enhance the strength of brickwork has not been researched and developed extensively as with concrete and its use is not extensive in the UK or throughout Europe. The concepts of design are very similar to those adopted for reinforced concrete but unlike concrete, brickwork is neither isotropic nor homogeneous, nor are the physical characteristics e.g. shrinkage, expansion etc. the same and care must be taken when comparing the two.

The processes of reinforcing and pre-stressing brickwork are generally straightforward and in most cases involve less effort than in concrete. In essence these involve taking advantage of the gaps which can be created using specific bonding patterns in which to place the reinforcement and providing anchorage plates where required as shown in Figures 1.40 and 1.41.

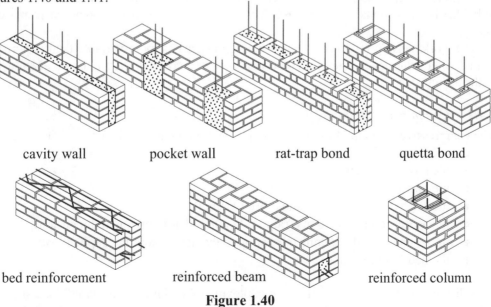

cavity wall   pocket wall   rat-trap bond   quetta bond

bed reinforcement   reinforced beam   reinforced column

Figure 1.40

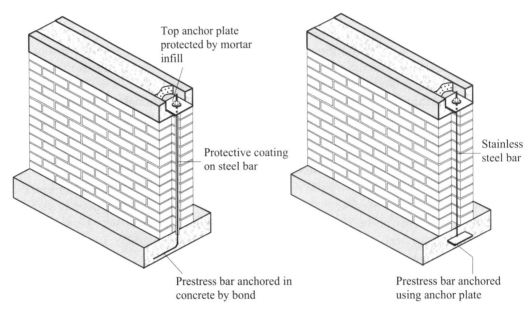

Figure 1.41

1.4   Material Properties

Masonry is a non-homogeneous, non-isotropic composite material which exists in many forms comprising units of varying shape, size and physical characteristics. The parameters which are most significant when considering structural design relate to strength and elastic properties, e.g. compressive, flexural and shear strengths, modulus of elasticity, coefficient of friction, creep, moisture movement and thermal expansion; these are discussed individually in sections 1.4.1 to 1.4.6. Tensile strength is generally ignored in masonry design.

   The workmanship involved in constructing masonry is more variable than is normally found when using most other structural materials and consideration must be given to this at the design stage.

1.4.1   Compressive Strength

The compressive strength of masonry is dependent on numerous factors such as:

♦   the mortar strength,
♦   the unit strength,
♦   the relative values of unit and mortar strength,
♦   the aspect ratio of the units (ratio of height to least horizontal dimension),
♦   the orientation of the units in relation to the direction of the applied load,
♦   the bed-joint thickness.

This list gives an indication of the complexity of making an accurate assessment of masonry strength. Unit strengths and masonry strengths are given in BS 5628 : Part 1:1992 in Figure 1(a) to 1(d) and Tables 2(a) to 2(d). These values are derived from research data carried out on individual units, small wall units (wallettes) and full-scale testing of storey height walls. The tabulated values are intended for use with masonry in which the structural units are laid on their normal bed faces in the attitude in which their compressive strengths are determined and in which they are normally loaded. Variations to this can be accommodated and testing procedures are specified in Appendix A of the code for the experimental determination of the characteristic compressive strength of masonry. Full-scale testing of storey height panels is considered to give the most accurate estimate of potential strength of masonry walls.

The failure mode of masonry in compression is usually one in which a tensile crack propagates through the units and the mortar in the direction of the applied load as shown in Figure 1.42. This crack is caused by secondary tensile stresses resulting from the restrained deformation of the mortar in the bed joints of the brickwork; see Hendry et al. (43) for a detailed explanation of the failure mechanism.

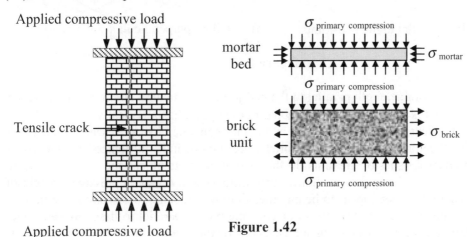

Figure 1.42

The tensile stresses inducing the crack are developed at the mortar-unit interfaces and are due to the restrained deformation of the mortar. In most cases masonry strength is considerably less than the strength of the individual units (see Table 2 of the code), it can, however, be considerably higher than the mortar strength. The apparent enhancement in the strength of the mortar is due to the biaxial or triaxial state of stress imposed on the mortar when it is acting compositely with the units.

The strength of masonry comprising low strength units with a high aspect ratio (see Table 2(d) of the code) and low strength mortars can be equal to the unit strengths. This difference in behaviour can be attributed to the combined low strengths and fewer bed-joints per unit height producing a more homogeneous composite.

1.4.2 *Flexural Strength*

The non-isotropic nature of masonry results in two principal modes of flexural failure which must be considered:

- ♦ failure parallel to the bed-joints, and

♦   failure perpendicular to the bed-joints.

As shown in Figures 1.43 (a) and (b).

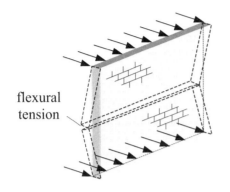
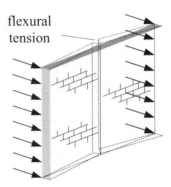

(a)   failure parallel to the bed-joints   (b)   failure perpendicular to the bed-joints

Figure 1.43

The ratio of (flexural strength parallel to the bed-joints) to (flexural strength perpendicular to the bed-joints) is known as the orthogonal ratio (μ) and has a typical value of 0.33 for clay, calcium silicate and concrete bricks and 0.6 for concrete blocks. Research indications are that the flexural strengths of clay bricks are significantly influenced by the water absorption characteristics of the units. In the case of concrete blocks the flexural strength *perpendicular* to the bed joints is significantly influenced by the compressive strength of the units. There does not appear to be any meaningful correlation between the strength of calcium silicate bricks, concrete bricks or concrete blocks parallel to the bed-joints, with any standard physical property. In all case the flexural strength in both directions is dependent on the strength of the mortar used and in particular the adhesion between the units and the mortar. The adhesion is very variable and research has shown it to be dependent on properties such as the pore structure of the units and mortar, the grading of the mortar sand and moisture content of the mortar at the time of laying.

These consideration have been included during the development of the values given in BS 5628 : Part 1. Table 3, *Characteristic flexural strength of masonry.*

1.4.3   Tensile Strength

As mentioned previously, the tensile strength of masonry is generally ignored in design. However, in Clause 24.1 the code indicates that a designer is permitted to assume 50% of the flexural strength values given in Table 3 when considering direct tension induced by suction forces arising from wind loads on roof structures, or by the probable effects of misuse or accidental damage.

Note: *In no circumstances may the combined flexural and direct tensile stresses exceed the values given in Table 3.*

1.4.4 Shear Strength

The shear strength of masonry is important when considering wall panels subject to lateral forces and structural forms such as diaphragm and fin walls where there is the possibility of vertical shear failure between the transverse ribs and flanges during bending.

Shear failure is most likely to be due to in-plane horizontal shear forces, particularly at the level of damp-proof courses.

The characteristic shear strength is dependent on the mortar strength and any precompression which exists. In BS 5628 : Part 1 : 1992, Clause 25 linear relationships are given for the characteristic shear strength and the precompression as follows:

(a)   Shear in a horizontal direction in a horizontal plane

Mortar Designations (i) and (ii)

$$f_v = 0.35 + 0.6g_A$$
$$\leq 1.75 \text{ N/mm}^2$$

Mortar Designations (iii) and (iv)

$$f_v = 0.15 + 0.6gA$$
$$\leq 1.4 \text{ N/mm}^2$$

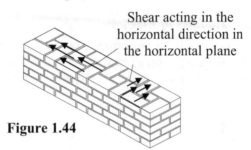

Shear acting in the horizontal direction in the horizontal plane

Figure 1.44

where:

f_v is the characteristic shear strength in the horizontal direction in the horizontal plane (see Figure 2 of the code),

g_A is the design vertical load per unit area of wall cross-section due to the vertical loads calculated for the appropriate loading conditions specified in Clause 22.

(b)   Shear in bonded masonry in the vertical direction in the vertical plane

For brick:

Mortar Designations (i) and (ii)

$$f_v = 0.7 \text{ N/mm}^2$$

Mortar Designations (iii) and (iv)

$$f_v = 0.5 \text{ N/mm}^2$$

For dense aggregate solid concrete block with a minimum strength of 7 N/mm² :

Mortar Designations (i) and (ii)

$$f_v = 0.7 \text{ N/mm}^2$$

Mortar Designations (iii) and (iv)

$$f_v = 0.5 \text{ N/mm}^2$$

Shear acting in the vertical direction in the vertical plane

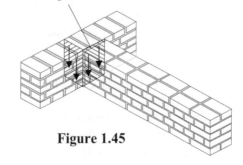

Figure 1.45

The design shear strength is obtained by applying the partial safety factor 'γ_{mv}' (see section 1.5), given in Clause 27.4 of the code and hence:

$$\text{Design shear strength of masonry} = \frac{f_v}{\gamma_{mv}}$$

where γ_{mv} = 2.5   when mortar not weaker than designation (iv) is used, and
           = 1.25 when considering the probable effects of misuse or accidental damage.

In the case of wall panels which are subject to lateral loads and restrained by concrete supports, the shear forces can be transmitted by metal wall ties. The characteristic shear strength of various types of tie which are engaged in dovetail slots in structural concrete are given in Table 8 of the code. As with masonry, the design values can be obtained by applying a partial safety factor 'γ_m' as given in Clause 27.5, i.e.
γ_m = 3.0 under normal conditions and
     = 1.5 when considering the probable effects of misuse or accidental damage.

1.4.5   Modulus of Elasticity

Since masonry is an anisotropic, composite material the value of elastic modulus, E_m, is variable depending on several factors such as materials used, direction and type of loading etc. A typical stress–strain curve is shown in Figure 1.46.

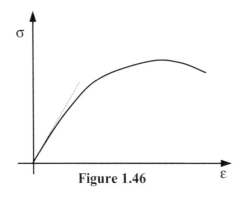

Figure 1.46

The actual value of E_m varies considerably and is approximated in BS 5628 : **Part 2** : 2000, Clause 7.4.1.7 as being equal to 0.9 f_k kN/mm^2. In practice this may range from 0.5 f_k to 2.0 f_k, the value given, however, is sufficiently accurate for design purposes.
In the long-term the value of elastic modulus allowing for creep and shrinkage may be taken as:

E_m = 0.45f_k kN/mm^2 for clay and dense aggregate concrete masonry
and
E_m = 0.3f_k kN/mm^2 for calcium silicate, autoclaved aerated concrete (a.a.c.) and lightweight concrete masonry

as indicated in Appendix C of Part 2 of the code.

In service masonry stresses are normally very low when compared with the ultimate value and consequently the use of linear elastic analysis techniques to determine structural deformations is acceptable.

1.4.6   Coefficient of Friction

The value of the coefficient of friction between clean concrete and masonry faces is given in Clause 26 as 0.6.

1.4.7 *Creep, Moisture Movement and Thermal Expansion*

The effects of creep, shrinkage, moisture and thermal movement are all significant particularly when considering the design of prestressed masonry. In each case the loss of prestressing force can occur and at low levels of strain in the tendon the effects of prestress can be eliminated. They can all induce fine cracking and opening up of joints and may require the provision of movement joints.

The creep characteristics can be estimated as indicated in BS 5628 : **Part 2** : 2000, Clause 9.4.2.5, i.e. creep is numerically equal to 1.5 × the elastic deformation of the masonry for fired-clay or calcium silicate bricks and 3.0 × the elastic deformation of the masonry for dense aggregate concrete blocks.

The moisture movement of masonry, [expansion (+) or contraction (–)], can be estimated using the values of shrinkage strain according to Clause 9.4.2.4 of BS 5628 : Part 2 : 2000, where shrinkage/expansion strain is equal to:

$$\varepsilon = -500 \times 10^{-6} \quad (-0.5 \text{ mm/m}) \qquad \text{for calcium silicate and concrete masonry.}$$

The effect of moisture expansion of fired-clay masonry on the force in the tendons of prestressed masonry can be ignored.

As indicated in Clause 9.4.2.8 consideration should be given to differential thermal movement between masonry and the prestressing tendons, especially where tendon stresses are low.

1.5 Limit State Design

The design code BS 5628 like most other structural design codes is based on the '*Limit State Design*' philosophy. The aim of design is to produce a structure which:

- ♦ is stable and possesses an adequate margin of safety against collapse,
- ♦ is safe and reliable in service,
- ♦ is economical to build and maintain,
- ♦ satisfactorily performs its intended function and
- ♦ is sufficiently robust such that damage to an extent disproportionate to the original cause will not occur.

The limit state design philosophy, which was formulated for reinforced concrete design in Russia during the 1930s, achieves these objectives by considering two 'types' of limit state under which a structure may become unfit for its intended purpose, they are;

1. the *Serviceability Limit State* in which a condition, e.g. deflection, vibration or cracking occurs to an extent, which is unacceptable to the owner, occupier, client etc. and

2. the ***Ultimate Limit State*** in which the structure, or some part of it, is unsafe for its intended purpose e.g. compressive, tensile, shear or flexural failure or instability leading to partial or total collapse.

The basis of the approach is statistical and lies in assessing the probability of reaching a given limit state and deciding upon an acceptable level of that probability for design purposes. The method in most codes, including BS 5628, is based on the use of ***characteristic values*** and ***partial safety factors***.

1.5.1   Partial Safety Factors: the use of partial safety factors, which are applied separately to individual parameters, enables the degree of risk for each one to be varied e.g. reflecting the differing degrees of control which are possible in the manufacturing process of building structural materials/units (i.e. mortar and individual bricks) and the construction of masonry (i.e. composite of mortar and units).

1.5.2   Characteristic Values: the use of characteristic values enables the statistical variability of various parameters e.g. material strength, different load types etc. to be incorporated in an assessment of the acceptable probability that the value of the parameter will be exceeded during the life of a structure. The term 'characteristic' in current design codes normally refers to a value of such magnitude that statistically only a 5% probability exists of its being exceeded.

In the design process the characteristic loads are multiplied by the partial safety factors to obtain the design values of design effects such as axial or flexural stress, and the design strengths are obtained by dividing the characteristic strengths by appropriate partial safety factors for materials. To ensure an adequate margin of safety the following must be satisfied:

$$\textbf{Design strength} \quad \geq \quad \textbf{Design load effects}$$

e.g.

$$\frac{f_k}{\gamma_m} \geq [(\text{stress due to } G_k \times \gamma_{f\,dead}) + (\text{stress due to } Q_k \times \gamma_{f\,imposed})]$$

where:
f_k is the characteristic compressive strength of masonry,
γ_m is the partial safety factor for materials (masonry),
G_k is the characteristic dead load,
Q_k is the characteristic imposed load,
$\gamma_{f\,dead}$ partial safety factor for dead loads,
$\gamma_{f\,imposed}$ is the partial safety factor for imposed loads.

The limit state philosophy can be expressed with reference to frequency distribution curves for design strengths and design effects as shown in Figure 1.47.

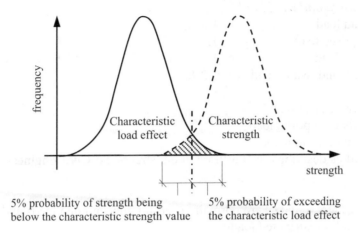

Figure 1.47

The shaded area represents the probability of failure, i.e. the level of design load effect which can be expected to be exceeded by 5% and the level of design strength which 5% of samples can be expected to fall below. The point of intersection of these two distribution curves represents the ultimate limit state, i.e. the design strength equals the design load effects.

The partial safety factors represent the uncertainty in the characteristic values. The lack of detailed statistical data on all of the parameters considered in design and complexity of the statistical analysis has resulted in the use of a more subjective assessment of the values of partial safety factors than is mathematically consistent with the philosophy.

In Clause 22 of BS 5628 : Part 1 : 1992 the following values for the partial safety factors (γ_f) applied to loads are given as:

(a) *Dead and imposed load*
   design dead load $= 0.9\,G_k$ or $1.4\,G_k$
   design imposed load$= 1.6Q_k$
   design earth and
   water load $= 1.4\,E_n$

(b) *Dead and wind load*
   design dead load $= 0.9\,G_k$ or $1.4\,G_k$
   design wind load $= 1.4\,W_k$ or $0.015W_k$
                whichever is the larger
   design earth and
   water load $= 1.4\,E_n$

In the particular case of freestanding walls and laterally loaded wall panels, whose removal would in no way affect the stability of the remaining structure, γ_f applied on the wind load may be taken as 1.2.

(c) *Dead, imposed and wind load*
    design dead load $= 1.2\,G_k$
    design imposed load $= 1.2\,Q_k$
    design wind load $= 1.2\,W_k$ or $0.015G_k$ whichever is the larger
    design earth and water load $= 1.2\,E_n$

where:
G_k is the characteristic dead load,
Q_k is the characteristic imposed load,
W_k is the characteristic wind load,
E_n is the earth load as described in 'Earth retaining structures' Civil Engineering Code of Practice No.2.

Note: Upper case letters (e.g. G_k) are normally used for concentrated loads and lower case letters (e.g. g_k) used for distributed loads.

1.6   Review Problems

1.1   Explain the difference between the terms: engineering brick, common brick, facing brick and special brick.
       (see section 1.1)

1.2   Identify the seven types of structural unit referred to in BS 5628.
       (see section 1.2)

1.3   State six criteria which may influence the choice of unit selected for a particular project.
       (see section 1.2)

1.4   Explain the difference between '*work sizes*' and '*co-ordinating sizes*' when designing brickwork.
       (see section 1.2)

1.5   Explain the purpose of mortar in masonry construction.
       (see section 1.2.2)

1.6   Explain why lime is sometimes used in mortar.
       (see section 1.2.2)

1.7   Explain the difference between mortars containing lime and lime mortars.
       (see section 1.2.2)

1.8   Explain the potential problems when introducing plasticisers into a mortar mix.
       (see section 1.2.2)

1.9   Explain the purpose of bonding in brickwork.
       (see section 1.2.3)

1.10   Explain the difference between '*jointing*' and '*pointing*' of brickwork.
       (see section 1.2.4)

1.11   Identify four types of joint finish.
       (see section 1.2.4)

1.12   Identify the most effective type(s) of joint finish to resist rain penetration.
       (see section 1.2.4)

1.13   State the purpose of a damp-proof course and identify the material
       properties required of it.
       (see section 1.2.5)

1.14   Explain the term '*rendering*' and the advantages of using a weak cement :
       lime based mix when carrying this out.
       (see section 1.2.6)

1.15   Explain the purpose of wall ties in cavity wall construction.
       (see sections 1.2.7 and 1.3.4)

1.16   Explain the purpose of a bitumen (or alternative) coating on an ashlar
       wall.
       (see section 1.3.2)

1.17   Explain the difference between a faced wall and a veneered wall.
       (see section 1.3.3)

1.18   Identify six factors which influence the compressive strength of masonry.
       (see section 1.4.1)

1.19   Define the orthogonal ratio with respect to flexural strength.
       (see section 1.4.2)

1.20   Explain, in general terms, the concept of limit state design.
       (see section 1.5)

2. Axially Loaded Walls

Objective: *'To illustrate the requirements for the limit-state design of axially loaded walls subject to concentgric or eccentric loads.'*

2.1   Introduction

Load-bearing walls resisting primarily vertical, in-plane loading, are often referred to as *'axially loaded walls'* whilst wall panels resisting wind loading are known as *'laterally loaded wall panels.'* The most commonly used types of axially loaded elements are:

♦ single-leaf (solid) walls,
♦ cavity walls,
♦ walls stiffened with piers   and
♦ columns,

as indicated in Figure 2.1 in this text and in Figure 3 of BS 5628 : Part 1 : 1992.

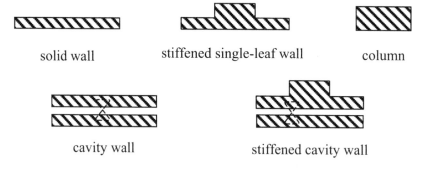

solid wall   stiffened single-leaf wall   column

cavity wall   stiffened cavity wall

Figure 2.1

Each of these types of element may be subject to *'concentric'* axial loading inducing compression only or *'eccentric'* axial loading resulting in combined compressive and bending forces.

The magnitude of loading which can be sustained by a load-bearing wall or column is dependent on a number of factors such as the:

♦ characteristic compressive strength of masonry, i.e. combined units and mortar (f_k),
♦ partial safety factor for the material strength (γ_m),
♦ plan area,

- thickness of the wall (t),
- slenderness of the element (h_{eff}/t_{eff}),
- eccentricity of the applied load (e_x),
- combined slenderness and eccentricity,
- eccentricities about both axes of a column,
- type of structural unit, and
- laying of structural units.

Each of these factors is discussed separately in sections 2.1.1 to 2.1.10.

2.1.1 *Characteristic Compressive Strength of Masonry (f_k)*

The characteristic compressive strength of masonry is given in Figures 1(a) to 1(d) and in Tables 2(a) to 2(d) of the code for a variety of different types of structural unit including standard format bricks, solid and hollow concrete blocks. Alternatively, as indicated in Clause 2.3.1, the value of f_k of any masonry '....*may be determined by tests on wall specimens, following the procedure laid down in A.2.*', where A.2. refers to Appendix A.2. in BS 5628 : Part 1 : 1992. Figure 1(a) and Table 2(a) from the code is reproduced in this text in Figure 2.2 and Figure 2.3 respectively.

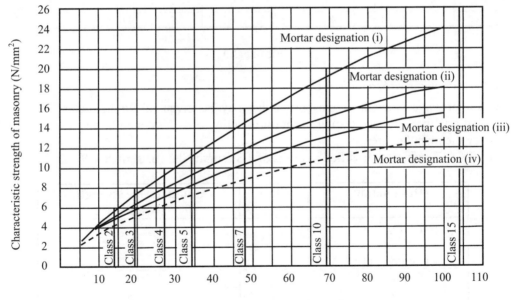

Figure 2.2

The value of f_k for any known combination of mortar designation and unit compressive strength can be determined from the tables. Linear interpolation within the tables is permitted, facilitated by the graphs given in Figures 1(a) to 1(d) of the code. The desired compressive strength of units and the mortar designation is normally specified by the designer.

Table 2(a) applies to masonry built with standard format bricks complying with the requirements of :

♦ BS 187 : 1987   *Specification for Calcium Silicate Bricks,*
♦ BS 6073 : Part 2 : 1981   *Specification for Precast Concrete Masonry Units* or
♦ BS 3921 : 1985   *Specification for Clay Bricks*

Table 2. Characteristic compressive strength of Masonry, f_k in N/mm^2

(a) Constructed with standard format bricks

Mortar desig-nation	Compressive strength of unit (N/mm^2)								
	5	10	15	20	27.5	35	50	70	100
(i)	2.5	4.4	6.0	7.4	9.2	11.4	15.0	19.2	24.0
(ii)	2.5	4.2	5.3	6.4	7.9	9.4	12.2	15.1	18.2
(iii)	2.5	4.1	5.0	5.8	7.1	8.5	10.6	13.1	15.5
(iv)	2.2	3.5	4.4	5.2	6.2	7.3	9.0	10.8	12.7

Figure 2.3

2.1.1.1 Compressive Strength of Unit

The compressive strength of masonry units may be given in N/mm^2 determined from standard quality assurance testing during production or be assigned to a specific strength class by the manufacturer.

In BS 187, units are classified in terms of a Strength Class as shown in Table 2.1.

Designation	Class	Compressive strength (N/mm^2)
Facing brick or Common brick	2	
Facing brick or Load-bearing brick	3	20.5
	4	27.5
	5	34.5
	6	41.5
	7	48.5

Table 2.1

Manufacturers are required to identify their products on relevant documentation, providing the following information:

- the name, trade mark or other means of identification of the manufacturer,
- the strength class of brick as designated in Table 2 of the code (Table 2.1 in this text),
- the work size, length, width and height, and whether with or without a frog,
- the number of the British Standard used i.e. BS 187.

In addition to this information some manufacturers colour code their bricks to indicate the Strength Class. The coding system specified in Clause 11 of BS 187 is as given in Table 2.2.

Strength Class	Mean compressive strength not less than (N/mm²)	Colour
2		–
3	20.5	Black
4	27.5	Red
5	34.5	Yellow
6	41.5	Blue
7	48.5	Green

Table 2.2

In BS 6073 : Part 2 : 1981, unit strengths are specified as shown in Table 2.3. The graphs in Figures 1(a) to 1(d) in BS 5628 : Part 1 : 1992 can be used for interpolation between these compressive strengths to determine the characteristic strength of the masonry.

Blocks (N/mm²)	Bricks (N/mm²)
2.8	7.0
3.5	10.0
5.0	15.0
7.0	20.0
10.0	30.0
15.0	40.0
20.0	
35.0	

Table 2.3

In Table 4 of BS 3921 : 1985 the classification of bricks by compressive strength and water absorption is given as shown in Table 2.4.

Note: It is important to note that there is no direct relationship between compressive strength and water absorption as given in this table, and durability.

Table 4. Classification of bricks by compressive strength and water absorption		
Class	**Compressive strength** (see 2.1)	**Water absorption** (see 2.2)
	N/mm^2	% by mass
Engineering A	≥ 70	≤ 4.5
Engineering B	≥ 50	≤ 7.0
Damp-proof course 1	≥ 5	≤ 4.5
Damp-proof course 2	≥ 5	≤ 7.0
All others	≥ 5	No limits
NOTE 1. There is no direct relationship between compressive strength and water absorption as given in this table and durability. NOTE 2. Damp-proof course 1 bricks are recommended for use in buildings whilst damp-proof course 2 bricks are recommended for use in external works (see table 13 of BS 5628 : Part 3: 1985).		

Table 2.4

2.1.1.2 Compressive Strength of Mortar

The most appropriate mortar for any particular application is dependent on a number of factors such as strength, durability and resistance to frost attack. In Table 13 of BS 5628 : Part 3 : 1985, guidance is given on the choice of masonry units and mortar designations most appropriate for particular situations regarding durability. The following categories are included:

- work below or near external ground level,
- damp-proof courses,
- unrendered external walls (other than chimneys, cappings, copings, parapets and sills),
- rendered external walls (other than chimneys, cappings, copings, parapets and sills),
- internal walls and inner leaves of cavity walls,
- unrendered parapets (other than cappings and copings),
- rendered parapets (other than cappings and copings),
- chimneys,
- cappings, copings and sills,
- freestanding boundary and screen walls (other than cappings and copings),
- earth-retaining walls (other than cappings and copings).

The mortar designations given in Table 1 of BS 5628 : Part 1 :1992 (see Table 2.5), have been selected to provide the most suitable mortar which will be readily workable to enable the production of satisfactory work at an economic rate and provide adequate durability.

The mortars are designated (i) to (iv), (i) being the strongest and most durable. Careful consideration is required when selecting a combination of structural unit and mortar designation; whilst the strongest mix may be suitable for clay brickwork in exposed situations, the inherent shrinkage in calcium silicate bricks caused by moisture movement can lead to cracking if strong, inflexible joints are present.

Table 1. Requirements for mortar						
		Mortar designa- tion	Type of mortar (proportion by volume)			Mean compressive strength at 28 days	
			Cement : lime : sand	Masonry cement : sand	Cement : sand with plasticizer	Prelim- inary (laboratory) tests	Site tests
Increasing strength ↑	Increasing ability to accommodate movement, e.g. due to settlement, temperature and moisture changes	(i) (ii) (iii) (iv)	1: 0 to ¼ : 3 1 : ½ : 4 to 4¼ 1 : 1 : 5 to 6 1 : 2 : 8 to 9	– 1 : 2 ½ to 3 ½ 1 : 4 to 5 1 : 5 ½ : 6 ½	– 1 : 3 to 4 1 : 5 to 6 1 : 7 to 8	N/mm^2 16.0 6.5 3.6 1.5	N/mm^2 11.0 4.5 2.5 1.0
Direction of change in properties is shown by the arrows			Increasing resistance to frost attack during construction ⟶ Improvement in bond and consequent resistance to rain penetration ⟵				

Table 2.5

The required 28 days mean compressive strength of the designated mortars, based on site tests, ranges from 11.0 N/mm^2 for designation (i) to 1.0 N/mm^2 for designation (iv). The site tests should be carried out according to the requirements of BS 4551 as indicated in Clause A.1.3 of BS 5628 : Part 1: 1992.

2.1.2   *Partial Safety Factor for Material Strength (γ_m)*

The design compressive strength of masonry is determined by dividing the characteristic strength (f_k), by a partial safety factor (γ_m), which is given in Table 4 of BS 5628 : Part 1 : 1992 and is shown in Table 2.6.

Table 4. Partial safety factors for material strength, γ_m			
		Category of construction control	
		Special	Normal
Category of manufacturing control of structural units	Special	2.5	3.1
	Normal	2.8	3.5

Table 2.6

The γ_m factor makes allowance for the inherent differences between the estimated strength characteristics as determined using laboratory tested masonry specimens and the actual strength of masonry constructed under site conditions and in addition allows for variations in the quality of materials produced during the manufacturing process.

The value of γ_m adopted is dependent on the degree of quality control practised by manufacturers and the standard of site supervision, testing and workmanship achieved during construction. There are two categories of control adopted in the code:

◆ normal control   and,
◆ special control.

2.1.2.1 Normal Control   (Clause 27.2.1.1 and Clause 27.2.2.1)

In **manufacturing**, normal control '*... should be assumed when the supplier is able to meet the requirements for compressive strength in the appropriate British Standard, but does not meet the requirements for the special category...*'.

In **construction**, normal control '*... should be assumed whenever the work is carried out following the recommendations for workmanship in section four of BS 5628 : Part 3 : 1985, or BS 5390, including appropriate supervision and inspection.*'

2.1.2.2 Special Control (Clause 27.2.1.2 and Clause 27.2.2.2)

In **manufacturing**, special control '*... may be assumed where the manufacturer:*
(a) *agrees to supply consignments of structural units to a specified strength limit, referred to as the acceptance limit, for compressive strength, such that the average compressive strength of a sample of structural units, taken from any consignment and tested in accordance with the appropriate British Standard specification, has a probability of not more than 2.5% of being below the acceptance limit, and*
(b) *operates a quality control scheme, the results of which can be made available to demonstrate to the satisfaction of the purchaser that the acceptance limit is consistently being met in practice, with the probability of failing to meet the limit being never greater than that stated in* **27.2.1.2 (a)**.'
In **construction**, special control   '*... may be assumed where the requirements of the*

normal category control are complied with and in addition:

    (a) the specification, supervision and control ensure that the construction is compatible with the use of the appropriate partial safety factors given in table 4;

    (b) preliminary compressive strength tests carried out on the mortar to be used, in accordance with A.1, indicate compliance with the strength requirements given in table 1 and regular testing of the mortar used on site, in accordance with A.1, shows that compliance with the strength requirements given in table 1 is being maintained.'

The value of γ_m in Table 4 applies to **compressive and flexural failure**. When considering the probable effects of misuse or accident, these values may be halved except where a member is deemed to be a *'protected member'*[1] as defined in Clause 37.1.1 of the code. As indicated in Clause 27.3, in circumstances where wall tests have been carried out in accordance with Appendix A of the code to determine the characteristic strengths, the γ_m values in Table 4 can be multiplied by 0.9.

When considering **shear failure**, the partial safety factor for masonry strength (γ_{mv}), should be taken as **2.5** as indicated in Clause 27.4.

The value of γ_m to be applied to the **strength of wall ties** is given in Clause 27.5 as **3.0**; as with compressive and flexural failure when considering the effects of misuse or accidental damage this value of γ_m can be halved.

2.1.3   *Plan Area (Clause 23.1.7)*

In Clause 23.1.7 allowance is made for the increased possibility of low strength units having an adverse effect on the strength of a wall or column with a small plan area, e.g. consider the two walls A and B shown in Figure 2.4.

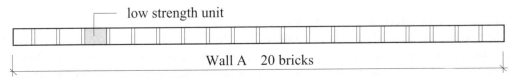

low strength unit

Wall A   20 bricks

The proportion of the cross-sectional area affected by the low strength unit in wall A is 5%

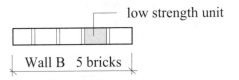

low strength unit

Wall B 5 bricks

The proportion of the cross-sectional area affected by the low strength unit in wall B is 20%

Figure 2.4

Although this effect is statistical it is essentially a geometrical effect on the compressive

---

[1] See Chapter 5 Section 5.9.1 for protected members.

strength. The allowance for this is made by multiplying the characteristic compressive strength, (f_k), by the factor:

$$(0.70 + 1.5A)$$

where:

A is the horizontal loaded cross-sectional area of the wall or column.

   This factor applies to all walls and columns where the cross-sectional area is less than 0.2 m². Clearly when $A = 0.2$ m² the factor is equal to 1.0.

2.1.4 Thickness of Wall 't' (Clause 23.1.2)

The compressive failure of walls occurs predominantly by the development of vertical cracks induced by the Poisson's ratio effect. The existence of vertical joints (as shown in Figure 2.5), reduces the resistance of brickwork to vertical cracking.

vertical joints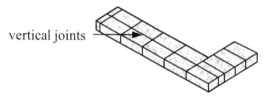

Figure 2.5

The indications from experimental evidence is that greater resistance to crack development is afforded by continuity in the cross-section such as occurs in narrow (half-brick) walls, than is the case when vertical joints are present. This increased resistance to compressive failure is reflected in a modification factor equal to 1.15 which is applied to the characteristic compressive strength (f_k), obtained from Table 2(a) for standard format bricks. The factor applies to single walls and loaded inner leafs of cavity walls as indicated in Clause 23.1.2 of the code. **(Note: this does not apply when both leaves are loaded.)**

2.1.4.1 Effective Thickness 't_{ef}'   (Clause 28.4)

The concept of effective thickness was introduced to determine the *'slenderness ratio'* when considering buckling due to compression. In Figure 3 of BS 5628 : Part 1 : 1992, values of effective thickness are given for various plan arrangements as shown in Figures 2.6 and 2.7. The modification to the actual thicknesses account for the stiffening effects of piers, intersecting and cavity walls which enhance the wall stability.

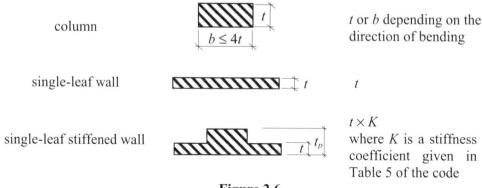

Figure 2.6

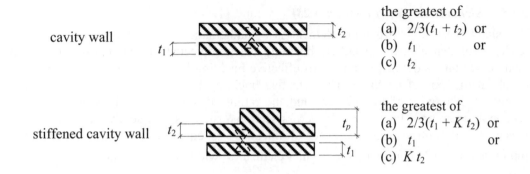

cavity wall

the greatest of
(a) $2/3(t_1 + t_2)$ or
(b) t_1 or
(c) t_2

stiffened cavity wall

the greatest of
(a) $2/3(t_1 + K\, t_2)$ or
(b) t_1 or
(c) $K\, t_2$

Figure 2.7

The stiffness coefficient K is given in Table 5 of the code as indicated in Figure 2.8.

Table 5. Stiffness coefficient for walls stiffened by piers

Ratio of pier spacing (centre to centre) to pier width	Ratio t_p/t of pier thickness to actual thickness of wall to which it is bonded		
	1	2	3
6	1.0	1.4	2.0
10	1.0	1.2	1.4
20	1.0	1.0	1.0
NOTE. Linear interpolation between the values given in table 5 is permissible, but not extrapolation outside the limits given.			

Figure 2.8

As indicated in Clause 28.4.2 where a wall is stiffened by intersecting walls, the value of K can be determined assuming that the intersecting walls are equivalent to piers of width equal to the thickness of the intersecting wall and of thickness equal to $3 \times$ the thickness of the stiffened wall as shown in Figure 2.9.

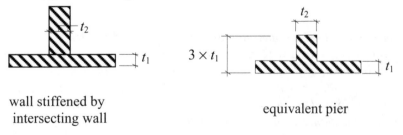

wall stiffened by
intersecting wall

equivalent pier

Figure 2.9

2.1.5   *Slenderness (h_{ef}/t_{ef}) (Clause 28)*

The slenderness of a structural element is a mathematical concept used to assess the tendency of that element to fail by buckling when subjected to compressive forces. In many cases this is defined as a ratio of effective buckling length (l_e) to radius of gyration (r) about the axis of buckling. The effective buckling length is related to the type and degree of end fixity of the element and the radius of gyration is related to the cross-sectional geometry. In rectangular cross-sections such as frequently encountered in masonry and timber design, the actual thickness is equal to the radius of gyration multiplied by $\sqrt{12}$; consider a rectangular element of width '*d*' and thickness '*t*':

$$r_{yy} = \sqrt{\frac{I_{yy}}{\text{Area}}} = \sqrt{\frac{dt^3}{12} \times \frac{1}{dt}} = \frac{t}{\sqrt{12}}$$

$$t = \sqrt{12}\, r_{yy} \qquad \therefore t \propto r_{yy}$$

Since '*t*' is directly proportional to '*r*' then slenderness can equally well be expressed in terms of thickness instead of radius of gyration.

In masonry design, in Clause 3.19 of BS 5628 : Part 1 : 1992 the slenderness ratio is defined as '*The ratio of effective height or length to the effective thickness.*'

The effective thickness as described in section 2.1.4.1 is a modification of the actual thickness to account for different plan layouts.

The slenderness ratio should **not exceed 27**, except in the case of walls less than 90 mm thick, in buildings more than two storeys, where it should **not exceed 20** (see Clause 28.1).

2.1.5.1 *Effective Height (Clause 28.3.1)*

Walls (Clause 28.3.1.1);
'*The effective height of a wall may be taken as:*
   (a) 0.75 times the clear distance between lateral supports which provide enhanced resistance to lateral movement, or
   (b) the clear distance between lateral supports which provide simple resistance to lateral movement.'

Columns (Clause 28.3.1.2);
'*The effective height of a column should be taken as the distance between lateral supports or twice the height of the column in respect of a direction in which lateral support is not provided.*'

Columns formed by adjacent openings in walls (Clause 28.3.1.3);
'*Where openings occur in a wall such that the masonry between any two opening is, by definition[2], a column, the effective height of the column should be taken as follows.*
   (a) Where an enhanced resistance to lateral movement of the wall containing the column is provided, the effective height should be taken as 0.75 times the distance between the supports plus 0.25 times the height of the taller of the two openings.

---
[2] Clause 3.7 An isolated vertical member whose width is not more than four times its thickness.

    (b) Where a simple resistance to lateral movement of the wall containing the column is provided, the effective height should be taken as the distance between the supports'

These conditions are illustrated in Figure 2.10.

Note: It is important to ensure that the ***column*** and not only the wall is provided with the assumed restraint condition; particularly when it extends to the level of the support.

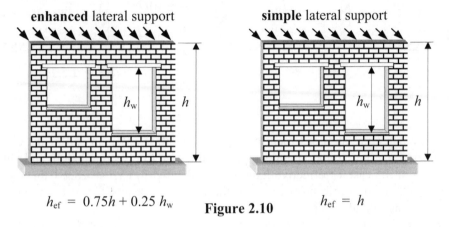

$$h_{ef} = 0.75h + 0.25\,h_w \qquad \textbf{Figure 2.10} \qquad h_{ef} = h$$

Piers (Clause 28.3.1.4);
'Where the thickness of a pier is not greater than 1.5 times the thickness of the wall of which it forms a part, it may be treated as a wall for effective height considerations; otherwise the pier should be treated as a column in the plane at right angles to the wall.

NOTE. The thickness of a pier, t_p, is the overall thickness of the wall or, when bonded into one leaf of a cavity wall, the thickness obtained by treating this leaf as an independent wall.'

2.1.5.2 Effective Length (Clause 28.3.2)
'The effective length of a wall may be taken as:

    *(a) 0.75 times the clear distance between vertical lateral supports or twice the distance between a support and a free edge, where lateral supports provide **enhanced resistance** to lateral movement,* (see Figure 2.11).

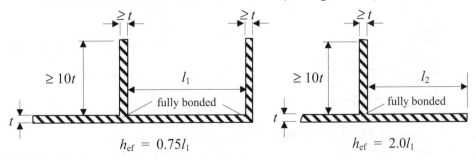

Figure 2.11

*(b) the clear distance between lateral supports or 2.5 times the distance between a support and a free edge where lateral supports provide **simple resistance** to lateral movement.'* (see Figure 2.12).

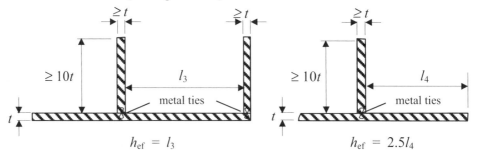

Figure 2.12

These definitions require an assessment of support conditions at each end of the element being considered, i.e. horizontal lateral support in the case of effective height and vertical lateral support in the case of effective length. As indicated in Clause 28.2.1 the supports should be capable of transmitting the sum of the following design forces to the principal elements providing lateral stability to the structure:

(a) 'the simple static reactions to the total applied design horizontal forces at the line of lateral support, and

(b) 2.5% of the total design vertical load that the wall or column is designed to carry at the line of the lateral support; the elements of construction that provide lateral stability to the structure as a whole need not be designed to support this force.'

Two types of horizontal and vertical lateral supports are defined in the code:

♦ simple supports   and
♦ enhanced supports.

2.1.5.3 Horizontal Simple Supports (Clause 28.2.2.1)

All types of floors and roofs which abut walls provide simple support if they are detailed as indicated in Appendix C of the code; a few typical examples are given in Figures 2.13 to 2.21.

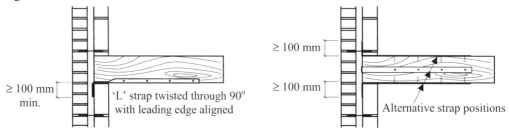

Figure 2.13   Timber floor bearing directly on to a wall

Note: In houses up to three storeys no straps are required, provided that the joist spacing is not greater than 1.2 m and joist bearing is 90 mm minimum.

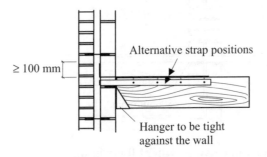

Figure 2.14   Timber floor using typical floor hanger

Note: In houses up to three storeys no straps are required, provided that the joist is effectively tied to the hanger.

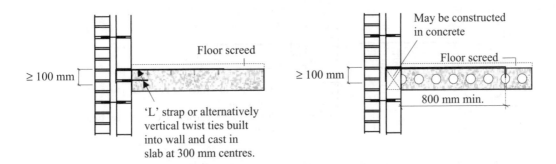

Figure 2.15   In-situ and precast concrete floor abutting external cavity wall

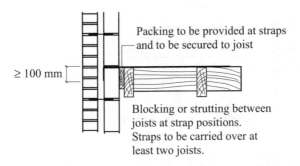

Figure 2.16   Timber floor abutting external cavity wall

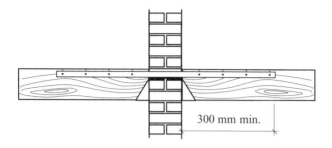

Figure 2.16   Timber floor using typical floor hanger

Note: In houses up to three storeys no straps are required, provided that the joist is effectively fixed to the hanger. Such fixing can be assumed if joist hangers as shown in Figures 14 and 15 of the code, are provided at no more than 2.0 m centres, with typical hangers in between.

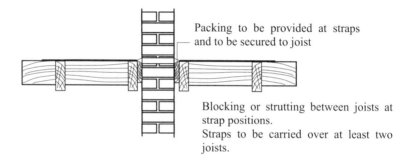

Figure 2.17   Timber floor abutting internal wall

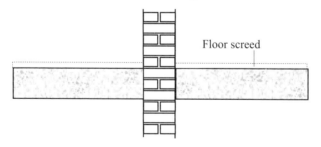

Figure 2.18   In-situ concrete floor abutting internal wall

2.1.5.4 Horizontal Enhanced Supports (Clause 28.2.2.2)

Enhanced lateral support can be assumed where:

  '*(a) floors and roofs of any form of construction span on to the wall or column from both sides at the same level;*

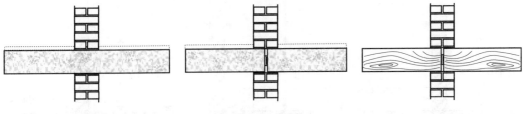

Figure 2.19

(b) *an in-situ concrete floor or roof, or a precast concrete floor or roof giving equivalent restraint, irrespective of the direction of span, has a bearing of at least one-half the thickness of the wall or inner leaf of a cavity wall or column on to which it spans but in no case less than 90 mm;*

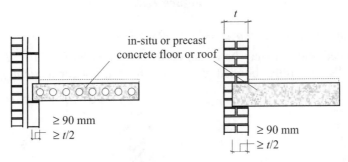

Figure 2.20

(c) *in the case of houses of not more than three storeys, a timber floor spans on to a wall from one side and has a bearing of not less than 90 mm.*

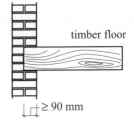

Figure 2.21

Preferably, columns should be provided with lateral support in both horizontal directions.'

2.1.5.5 Vertical Simple Supports (Clause 28.2.3.1)

Simple lateral support may be assumed where '... *an intersecting or return wall not less than the thickness of the supported wall or loadbearing leaf of a cavity wall extends from the intersection at least ten times the thickness of the supported wall or loadbearing leaf and is connected to it by metal anchors calculated in accordance with 28.2.1, (see 2.1.5.2), and evenly distributed throughout the height at not more than 300 mm centres.'*

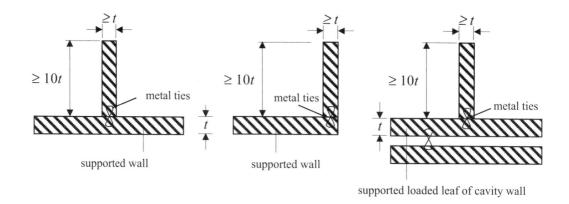

Figure 2.22

2.1.5.6 *Vertical Enhanced Supports (Clause 28.3.2)*

Vertical enhanced support may be assumed as the cases indicated in 2.1.5.5 where the supporting walls are fully bonded providing restraint against rotation as shown in Figure 2.23.

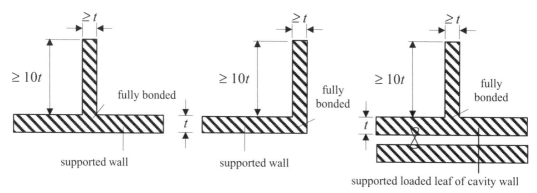

Figure 2.23

2.1.6   *Eccentricity of Applied Loading (Clauses 30, 31 and 32.1)*

In most cases the applied loading on a wall is not concentric. This is due to factors such as construction details, tolerances and non-uniformity of materials. The resultant eccentricity may be in the plane of the wall as in the case of masonry shear-walls resisting combined lateral wind loading and vertical floor/roof loading or, perpendicular to the plane of the wall as in the case of walls supporting floor/roof slabs and/or beams spanning on to them.

2.1.6.1 *Eccentricity in the Plane of a Wall*

This type of eccentricity is considered in Clauses 30 and 32.1 and Figure 4 of the code. The load distribution along a length of wall subject to both lateral and vertical loading can be determined using statics by combining both axial and bending effects as shown in Figure 2.24.

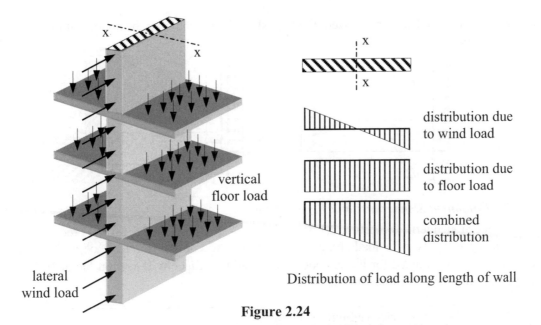

lateral wind load

vertical floor load

distribution due to wind load

distribution due to floor load

combined distribution

Distribution of load along length of wall

Figure 2.24

Standard methods of elastic analysis assuming rigid-floor plate deformation and consequently load distribution in proportion to the relative stiffness of walls, can be used to determine the lateral loading on any particular shear-wall within a building. If the plan layout of the walls in a building is asymmetric, it may be necessary to consider the effects of torsion in the analysis procedure to calculate the lateral load distribution.

2.1.6.2 *Eccentricity Perpendicular to the Plane of a Wall*

In Clause 31 of the code it is stated that '*Preferably, eccentricity of loading on walls and columns should be calculated but, at the discretion of the designer it may be assumed that.....*'. Whilst this calculation is possible (41), most engineers adopt the values suggested in the remainder of this clause, i.e.

♦ Where the load is transmitted to a wall by a single floor or roof it is assumed to act at one-third of the depth of the bearing area from the loaded face of the wall or loadbearing leaf.

♦ Where a uniform floor is continuous over a wall, each side of the floor may be taken as being supported individually on half the total bearing area.

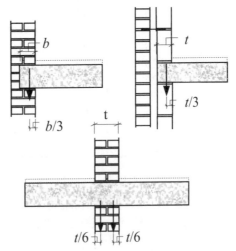

♦ Where joist hangers are used, the load should be assumed to be applied at the face of the wall.

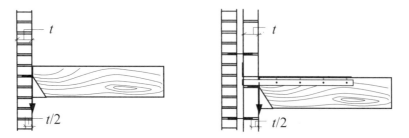

2.1.7   *Combined Slenderness and Eccentricity*

The effect of slenderness (section 2.1.5), and eccentricity (section 2.1.6), is to reduce the loadbearing capacity of a loaded wall or column. The combined effects of both these characteristics is allowed for in the code by evaluating a capacity reduction factor 'β' as shown in Figure 2.25 (Table 7 of the code).

Table 7. Capacity reduction factor, β				
Slender- ness ratio h_{ef}/t_{ef}	**Eccentricity at top of wall, e_x**			
	Up to 0.05t (see note 1)	0.1t	0.2t	0.3t
0	1.00	0.88	0.66	0.44
6	1.00	0.88	0.66	0.44
8	1.00	0.88	0.66	0.44
10	0.97	0.88	0.66	0.44
12	0.93	0.87	0.66	0.44
14	0.89	0.83	0.66	0.44
16	0.83	0.77	0.64	0.44
18	0.77	0.70	0.57	0.44
20	0.70	0.64	0.51	0.37
22	0.62	0.56	0.43	0.30
24	0.53	0.47	0.34	
26	0.45	0.38		
27	0.40	0.33		
NOTE 1. It is not necessary to consider the effects of eccentricities up to and including 0.05t. NOTE 2. Linear interpolation between eccentricities and slenderness ratios is permitted. NOTE 3. The derivation of β is given in Appendix B.				

Figure 2.25

The derivation of the capacity reduction factor 'β' is given in Appendix B of the code. The application of β is based on several assumptions:

- only braced vertical walls/columns are considered,
- additional moments caused by buckling effects are included,
- at failure, the stress distribution at the critical section can be represented by a rectangular stress block.

i)   *Only braced vertical walls/columns are considered:*
When side-sway of a wall is not permitted it can be assumed that the eccentricity of the applied load at the top of a wall varies from e_x at the top to zero at the bottom as shown in Figure 2.26.

Figure 2.26

ii)   *Additional moments caused by buckling effects:*
The tendency for a section to buckle is dependent on its slenderness. The lateral movement caused by this effect induces a secondary bending moment in the cross-section because of the additional eccentricity of the applied load. These secondary moments are allowed for in the code by increasing the value of eccentricity used in design and consequently reducing the β value. The additional eccentricity 'e_a' (see Figure 2.27) is assumed to vary linearly from zero at the top and bottom of a wall, to a value over the central fifth of the wall given by:

$$e_a = t \left[\frac{1}{2400} (h_{ef}/t_{ef})^2 - 0.015 \right]$$

where:
t   is the thickness of the wall or depth of the column,
t_{ef} is the effective thickness of the wall or column,
h_{ef} is the effective height of the wall or column.

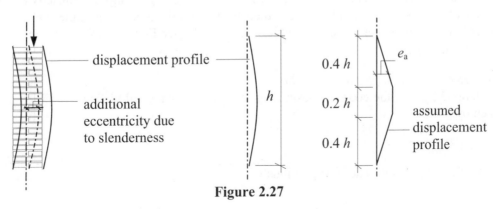

Figure 2.27

The addition of the eccentricities indicated in Figures 2.26 and 2.27 result in the total design eccentricity, e_t. Clearly the location of the maximum contribution from e_a is $0.4h$ below the top, i.e.

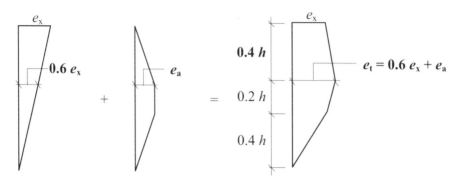

Figure 2.28

The value of e_t is calculated using $(0.6 \, e_x + e_a)$ as indicated in equation (2) of Appendix B in the code. In stocky (i.e. non-slender) walls the value of e_a may be less than $0.4e_x$ in which case the total design eccentricity e_t calculated using equation (2) will be smaller than e_x at the top of the wall. In such cases the maximum value, i.e. e_x, should be used for design (e.g. when $e_x = 10$ units and $e_a = 6$ units $e_t > e_x$ and when $e_x = 10$ units and $e_a = 3$ units $e_t < e_x$).

The equation used to calculate e_a has been derived on the basis of experimental data. The additional eccentricity becomes less significant as walls become less slender and less likely to buckle. A limiting value for slenderness, below which the additional eccentricity can be ignored, can be determined by equating e_a to zero; i.e.

$$e_a \quad = \quad t\left[\frac{1}{2400}\left(h_{ef}/t_{ef}\right)^2 - 0.015\right] \quad = \quad 0 \quad \therefore \quad \left[\frac{1}{2400}\left(h_{ef}/t_{ef}\right)^2 - 0.015\right] \quad = \quad 0$$

The limiting value of slenderness $= \; h_{ef}/t_{ef} \; = \; 6$

As a consequence of this no additional effects need be considered for walls/columns where the slenderness is less than or equal to 6. In addition when the design eccentricity e_m (the greater of e_t and e_x), is small, i.e. less than $0.05t$, its effects are negligible and can be ignored. These limits are evident in Table 7 of the code (see Figure 2.25), where the value of β is equal to 1.0.

iii)   *Rectangular stress block at failure*
In Clause 32.2.1 of the code the design vertical load resistance of walls per unit length is given by:

$$\frac{\beta \, t \, f_k}{\gamma_m}$$

where all symbols are as defined previously.

The design vertical load resistance of walls per unit length can also be expressed in terms of the design eccentricity e_m as indicated in Appendix B.

Consider the section of wall shown in Figure 2.29 in which the load is applied at a design eccentricity of e_m and the failure stress is assumed uniform and equal to $1.1 f_k/\gamma_m$. The 1.1 multiplying factor is a 10% adjustment in the stress to allow for correlation between the predicted failure loads and experimental evidence.

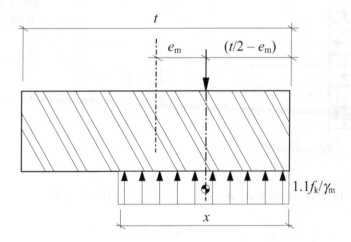

Figure 2.29

The line of action of the applied load must coincide with the centroid of the stress block. The length of the stress block can then be determined in terms of the design eccentricity. i.e.

length of stress block $\quad = \quad x \quad = \quad 2(t/2 - e_m) \quad = \quad (t - 2e_m)$
design load /unit length $\quad = \quad$ (area × stress) $\quad = \quad [(t - 2e_m) \times 1.0] \times [1.1 f_k/\gamma_m]$

This load must also be equal to the expression given in the code, i.e. $\dfrac{\beta\, t\, f_k}{\gamma_m}$

$$\therefore \quad \frac{\beta\, t\, f_k}{\gamma_m} \quad = \quad [(t - 2e_m) \times 1.0] \times [1.1 f_k/\gamma_m]$$

$$\therefore \quad \beta \quad = \quad 1.1[1 - (2e_m/t)]$$

This equation has been used to calculate the values given for β in Table 7 of the code.

Note: It is important to note that e_m is a function of e_x, the eccentricity at the top of the wall, and the design parameter used in Table 7 is e_x. In addition linear interpolation between eccentricities and slenderness values given in the code is permitted.

2.1.8   Eccentricities in Columns   (Clause 32.2.2)

Generally, the eccentricity of loading on a wall is relative to the axis parallel to the centre line as indicated in Figure 2.30(a). In columns, it is common for the applied loading to be eccentric relative to both the major and minor axes as indicated in Figure 2.30(b).

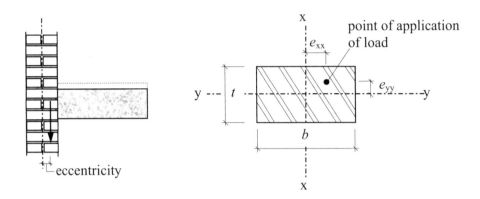

(a) vertical cross-section of wall   (b) plan of column

Figure 2.30

In Clause 32.2.2 the design vertical load resistance of rectangular columns is given by:

$$\frac{\beta\, bt\, f_k}{\gamma_m}$$

where all symbols are as defined previously.

The value of the capacity reduction factor β is dependent on the magnitudes of the eccentricities e_{xx} and e_{yy} of the applied load about the two principal axes. Four different cases are considered in the code:

(a) When the eccentricities about the major and minor axes at the top of the column are less than $0.05b$ and $0.05t$ Table 7 is used basing the **slenderness** ratio on an effective height h_{ef} about the **minor axis** and an effective thickness t_{ef} equal to t.

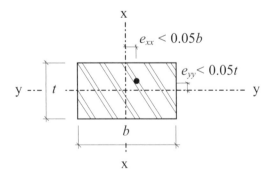

$$SR = \frac{h_{ef}\ (relative\ to\ the\ minor\ axis)}{t_{ef}\ (based\ on\ column\ thickness\ t)}$$

(b) When the eccentricities about the major and minor axes at the top of the column are less than $0.05b$ but greater than $0.05t$ respectively Table 7 is used basing the **slenderness** ratio on an effective height h_{ef} about the **minor axis** and an effective thickness t_{ef} equal to t and the **eccentricity** related to the **minor axis**.

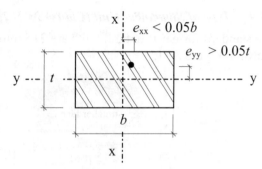

$$SR = \frac{h_{ef} \; (relative \; to \; the \; minor \; axis)}{t_{ef} \; (based \; on \; column \; thickness \; t)}$$

Eccentricity $e_x = e_{yy}$

(c) When the eccentricities about the major and minor axes at the top of the column are greater than $0.05b$ and less than $0.05t$ respectively Table 7 is used basing the **slenderness** ratio on an effective height h_{ef} about the **minor** axis and an effective thickness t_{ef} equal to t and the **eccentricity** related to the **major** axis. Alternatively Appendix B can be used considering slenderness and eccentricities about both axes.

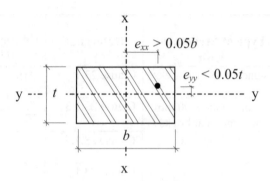

$$SR = \frac{h_{ef} \; (relative \; to \; the \; minor \; axis)}{t_{ef} \; (based \; on \; column \; thickness \; t)}$$

Eccentricity $e_x = e_{xx}$ or
Use Appendix B to evaluate β

(d) When the eccentricities about the major and minor axes at the top of the column are greater than $0.05b$ and $0.05t$ respectively **Appendix B** is used to determine β considering eccentricities and slenderness about both axes.

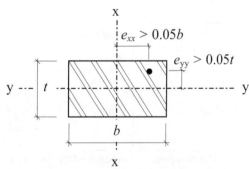

Use Appendix B to evaluate β considering both axes

2.1.9   Type of Structural Unit (Clause 23.1, 23.1.3 to 23.1.9)

The standard format structural units are 215 mm wide $\times$ 102.5 mm thick $\times$ 65 mm high as shown in Figure 2.31.

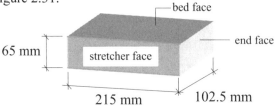

Figure 2.31

The characteristic strength f_k, of various types of masonry unit are specified in Tables 2(a) to 2(d) and Clauses 23.1.3 to 23.1.9 of the code; these are summarised in Figure 2.32.

Type of Structural Unit	Reference	Dimensions H-height L-least horizontal dimension	Characteristic Strength (f_k)
standard format bricks	Clause 23.1	215 mm $\times$ 102.5 mm $\times$ 65 mm	Table 2(a)
90 mm wide $\times$ 90 mm high modular bricks	Clause 23.1.3	thickness equal to width	Table 2(a) $\times$ 1.25
		other thicknesses	Table 2(a) $\times$ 1.10
wide bricks	Clause 23.1.4	H : L < 0.6	Testing according to A.2
blocks	Clause 23.1	H : L = 0.6	Table 2(b)
hollow blocks	Clause 23.1.5	0.6 < H : L < 2.0	Interpolation between Tables 2(b) & 2(c)
solid concrete blocks (i.e. no cavities)	Clause 23.1.6	0.6 < H : L < 2.0	Interpolation between Tables 2(b) & 2(d)
hollow blocks	Clause 23.1	2.0 < H : L < 4.0	Table 2(c)
hollow concrete blocks filled with concrete	Clause 23.1.7	see Note 1	Interpolation between Tables 2(b) & 2(d)
natural stone	Clause 23.1.8	Design on the basis of solid concrete blocks of equivalent compressive strength; see Clause 23.1.6	
random rubble	Clause 23.1.9	0.75 $\times$ natural stone masonry of similar materials. If built with lime-mortar use 0.5 $\times$ masonry in mortar designation (iv)	
Note 1: The compressive strength of the blocks should be based on net area and the 28-day cube strength of the concrete infill should not be less than this value.			

Figure 2.32

2.1.10 Laying of Structural Units (Clause 8)

Structural units are normally laid on their bed face and the strengths given in Tables 2(a) to 2(d) relate to this. Where units are laid on a different face, i.e. the stretcher or end face, their strength in this orientation should be determined by testing according to the requirements of BS 187, BS 6073 : Part 1 and BS 3921 as appropriate.

To develop the full potential strength of units with frogs it is necessary to ensure that a full and correct bedding of mortar is made and the frogs are filled. In circumstances where a lower frog cannot be filled or is only partially filled, tests should be carried according to BS 187 and BS 3921 to determine the masonry strength.

Note: When bed joints are raked out for pointing it is important to make allowance for the resulting loss of strength as indicated in Clause 8 of the code.

The design parameters discussed in section 2.1 are illustrated in Examples 2.1 to 2.15.

2.2   Example 2.1   Single-Leaf Masonry Wall 1

A 150 mm thick reinforced concrete roof slab is supported on two single-leaf walls as shown in Figure 2.33. Using the design data provided determine a suitable structural unit / mortar combination for the walls using:

i)   standard format bricks,
ii)   100 mm hollow blocks with a height-to-width ratio of 0.6,
iii)   90 mm × 90 mm modular bricks.

Design data:

Characteristic self-weight of concrete	24 kN/m^3
Assume the characteristic self-weight of all walls	2.0 kN/m^2
Characteristic imposed load on roof slab	1.5 kN/m^2
Category of manufacturing control	normal
Category of construction control	normal

Assume that the walls are part of a braced structure.

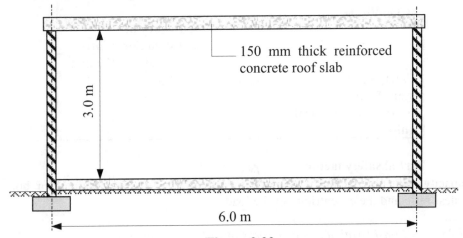

Figure 2.33

Solution:
Consider a 1.0 metre length of wall
(i) standard format bricks

Characteristic self-weight of all walls	$= (0.4 \times 3.0 \times 2.0)*$
	$= \textbf{2.4 kN/m}$
Characteristic dead load due to the self-weight of the slab	$= 24 \times (6.0 \times 1.0 \times 0.15)$
	$= 21.6$ kN/m length
	$= \textbf{10.8 kN/m per wall}$
Characteristic imposed load on the roof slab	$= (6.0 \times 1.0 \times 1.5)$
	$= 9.0$ kN/m
	$= \textbf{4.5 kN/m per wall}$

* **Note:** the critical section occurs at a distance of $0.4h$ down the wall, i.e. maximum eccentricity.

Clause 22(a)   *Partial Safety Factor for Loads (γ_f)*
Dead loads $\gamma_f = 1.4$
Imposed loads $\gamma_f = 1.6$
Design load/metre length of wall $= [(1.4 \, (2.4 + 10.8) + (1.6 \times 4.5)] = \textbf{25.68 kN}$

Clause 32.2.1   Design Vertical Load Resistance of Walls

Design vertical load resistance/unit length $= \dfrac{\beta t f_k}{\gamma_m}$ this value must be ≥ 22.32 kN

The required unknown is the characteristic compressive strength of the masonry f_k

$$f_k \geq \frac{22.32 \times \gamma_m}{\beta t}$$

Clause 27.3   *Partial Safety Factor for Material Strength (γ_m)*
                Category for manufacturing control is normal
                Category for construction control is normal
Table 4

<table>
<tr><td colspan="2" rowspan="2">Table 4. Partial safety factors for material strength, γ_m</td><td colspan="2">Category of construction control</td></tr>
<tr><td>Special</td><td>Normal</td></tr>
<tr><td rowspan="2">Category of manufacturing control of structural units</td><td>Special</td><td>2.5</td><td>3.1</td></tr>
<tr><td>Normal</td><td>2.8</td><td>3.5</td></tr>
</table>

Partial safety factor $\gamma_m = \textbf{3.5}$
The capacity reduction factor β is given in Table 7 and requires values for both the slenderness ratio and the eccentricity of the load.

Clause 28   *Consideration of slenderness of walls and columns*
                slenderness ratio (SR)$= h_{ef} / t_{ef} \quad \leq \quad 27$

Clause 28.2.2   Horizontal Lateral Support
Since this structure has a concrete roof with a bearing length of at least one-half the thickness of the wall, enhanced resistance to lateral movement can be assumed.

Clause 28.3.1   Effective Height
$$h_{ef} = 0.75 \times \text{clear distance between lateral supports}$$
$$= (0.75 \times 3000) = 2250 \text{ mm}$$

Clause 28.4.1   Effective Thickness
For single-leaf walls the effective thickness is equal to the actual thickness as indicated in Figure 3 of the code.
$$t_{ef} = 102.5 \text{ mm}$$
$$\text{SR} = \frac{2250}{102.5} \approx \mathbf{22} \quad < 27$$

Clause 31   Eccentricity Perpendicular to the Wall

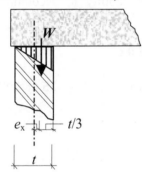

The load maybe assumed to act at an eccentricity equal to one-third of the depth of the bearing area from the loaded face of the wall.

$$\text{eccentricity } (e_x) = (1/2t - 1/3t) = t/6$$
$$e_x = \mathbf{0.167t}$$

Note that the eccentricities in Table 7 of the code are given in terms of the thickness *t*.

Table 7   Capacity Reduction Factor
Linear interpolation between slenderness and eccentricity values is permitted when using Table 7.

Table 7. Capacity reduction factor, β				
Slender-ness ratio h_{ef}/t_{ef}	**Eccentricity at top of wall, e_x**			
	Up to 0.05t (see note 1)	0.1t	0.2t	0.3t
0	1.00	0.88	0.66	0.44
18	0.77	0.70	0.57	0.44
20	0.70	0.64	0.51	0.37
22	0.62	0.56	0.43	0.30
24	0.53	0.47	0.34	

$$\beta = [0.56 - (0.13 \times 0.067/0.1)] = 0.47$$

Clause 32.2.1   Design Vertical Load Resistance

$$f_k \geq \frac{25.68 \times \gamma_m}{\beta t} = \frac{25.68 \times 3.5}{0.47 \times 102.5} = 1.87 \text{ N/mm}^2$$

Clause 23.1.2   Narrow Brick Walls
When using standard format bricks to construct a wall one brick (i.e. 102.5 mm) wide the values of f_k obtained from Table 2(a) can be multiplied by 1.15.

$$f_k \text{ required } = \frac{1.87}{1.15} = \textbf{1.62 N/mm}^2$$

It is evident from Table 2 of the code that any combination of brick strength and mortar designation will satisfy this requirement for f_k. Although this thickness of wall will satisfy the ultimate limit state requirement, consideration should also be given to other limit states e.g. resistance to rain penetration, frost attack and/or fire, ability to accommodate movement. Advice regarding these criteria can be found in BS 5628 : Part 3 : 1985.

(ii)   100 mm hollow blocks with a height-to-width ratio of 0.6
The applied load is the same as before:
    Design load/metre length of wall   = $[(1.4(2.4 + 10.8) + (1.6 \times 4.5)]$ = **25.68 kN**

Clause 27.3   Partial Safety Factor for Material Strength (γ_m)

$$\gamma_m = \textbf{3.5}$$

The capacity reduction factor β is given in Table 7 and requires values for both the slenderness ratio and the eccentricity of the load.

Clause 28   Consideration of Slenderness of Walls and Columns
                slenderness ratio (SR) = h_{ef}/t_{ef} $\leq$ 27

Clause 28.3.   The horizontal lateral support and effective height are the same as before:
                h_{ef} = 2250 mm

Clause 28.4.1   Effective Thickness
For single-leaf walls the effective thickness is equal to the actual thickness as indicated in Figure 3 of the code.
                t_{ef} = 100 mm
                $SR = \dfrac{2250}{100} = \textbf{22.5}$ < 27

Clause 31   Eccentricity Perpendicular to the wall
The eccentricity is the same as before: e_x = **0.167t**

Table 7   Capacity Reduction Factor

Table 7. Capacity reduction factor, β				
Slender-ness ratio h_{ef}/t_{ef}	**Eccentricity at top of wall,** e_x			
	Up to 0.05t (see note 1)	0.1t	0.2t	0.3t
0	1.00	0.88	0.66	0.44
20	0.70	0.64	0.51	0.37
22	0.62	0.56	0.43	0.3
24	0.53	0.47	0.34	
26	0.45	0.38		
27	0.4	0.33		

Linear interpolation between eccentricities of 0.1t and 0.2t is permissible:

$$\text{SR} = 22 \quad \beta_{ex=0.167} = [0.56 - (0.13 \times 0.067/0.1)] = 0.47$$
$$\text{SR} = 24 \quad \beta_{ex=0.167} = [0.47 - (0.13 \times 0.067/0.1)] = 0.38$$

SR	$\beta_{ex=0.167}$
22	0.47
24	0.38

$$\text{SR} = 22.5 \quad \beta = [0.47 - (0.09 \times 0.5/2.0)] = 0.45$$

Clause 32.2.1   Design Vertical Load Resistance $= \dfrac{\beta t f_k}{\gamma_m}$ /unit length

$$f_k \geq \frac{25.68 \times \gamma_m}{\beta t} = \frac{25.68 \times 3.5}{0.45 \times 100} = \textbf{2.0 N/mm}^2$$

Note: The narrow brick wall factor applies only to standard format bricks.

Table 2(b)   Characteristic Compressive Strength of Masonry
Any combination of unit strength $\geq$ 5.0 N/mm^2 and mortar designation (i) to (iv) will satisfy the masonry strength requirement.

(iii)   90 mm $\times$ 90 mm modular bricks
The applied load is the same as before:

Design load/metre length of wall $= [(1.4(2.4 + 10.8) + (1.6 \times 4.5)] = $ **25.68 kN**

Clause 27.3   *Partial Safety Factor for Material Strength* (γ_m)
$\gamma_m = \mathbf{3.5}$

Clause 28   *Consideration of Slenderness of Walls and Columns*
slenderness ratio (SR) = h_{ef}/t_{ef} $\leq$ 27

Clause 28.3   The horizontal lateral support and effective height are the same as before:
$h_{ef} = 2250$ mm

Clause 28.4.1   Effective Thickness
For single-leaf walls the effective thickness is equal to the actual thickness as indicated in Figure 3 of the code.

$t_{ef} = 90$ mm

$SR = \dfrac{2250}{90} = \mathbf{25}$ < 27

Clause 31   Eccentricity Perpendicular to the Wall
The eccentricity is the same as before: $e_x = \mathbf{0.167t}$

Table 7   Capacity Reduction Factor

Table 7. Capacity reduction factor, β

Slender-ness ratio h_{ef}/t_{ef}	Eccentricity at top of wall, e_x			
	Up to 0.05t (see note 1)	0.1t	0.2t	0.3t
0	1.00	0.88	0.66	0.44
20	0.70	0.64	0.51	0.37
22	0.62	0.56	0.43	0.3
24	0.53	0.47	0.34	
26	0.45	0.38		
27	0.4	0.33		

In Table 7 no values of β corresponding with slenderness ratios > 26 and eccentricities > 0.1t have been given, i.e. in the bottom right-hand corner. In circumstances where high eccentricities and high slenderness ratios exist the capacity reduction factor can be evaluated using the equations in Appendix B. If possible, it is advisable to avoid this situation by using higher effective thicknesses and/or specifying details which will reduce the eccentricity.

In this instance the value of β has been calculated to illustrate the procedure when using Appendix B.

Appendix B:

Equation 4 $\quad \beta = 1.1[1 - (2e_m/t)]$

where e_m is the larger of e_x and e_t and $e_t = (0.6e_x + e_a)$

Equation 1 $\quad e_a = t\left[\dfrac{1}{2400}\left(h_{ef}/t_{ef}\right)^2 - 0.015\right]$

$$e_x = 0.167t; \qquad \text{slenderness} = h_{ef}/t_{ef} = 25$$

$$e_a = t\left[\dfrac{1}{2400}(25)^2 - 0.015\right] = 0.245t$$

$$e_t = (0.6e_x + e_a) = [(0.6 \times 0.167t) + 0.245t] = 0.345t$$

$$e_m \geq e_x$$
$$\geq e_t \qquad\qquad \therefore \quad e_m = 0.345t$$

Equation 4 $\quad \beta = 1.1[1 - (2 \times 0.345t/t)] = 0.34$

Clause 32.2.1   Design Vertical Load Resistance $= \dfrac{\beta t f_k}{\gamma_m}$

$$f_k \geq \dfrac{25.68 \times \gamma_m}{\beta t} = \dfrac{25.68 \times 3.5}{0.34 \times 90} = \mathbf{2.94 \ N/mm^2}$$

Clause 23.1.3   Walls Constructed in Modular Bricks

When using 90 mm wide × 90 mm high modular bricks in single-leaf masonry 90 mm thick, f_k from Table 2(a) can be multiplied by 1.25 (1.1 for any other thickness).

$$f_k \text{ required} = \dfrac{2.94}{1.25} = \mathbf{2.35 \ N/mm^2}$$

Table 2(a)   Characteristic Compressive Strength of Masonry

Any combination of unit strength and mortar designation types (i), (ii) or (iii) will satisfy the masonry strength requirement.

2.3   Example 2.2   Single-Leaf Masonry Wall 2

A steel beam is supported by two masonry walls and carries a mid-span concentrated load from a stanchion in addition to a uniformly distributed load as shown in Figure 2.34. Using the data given, select an appropriate concrete block strength assuming a 1:3, masonry cement:sand mortar is to be used.

Design data:

Assume the characteristic self-weight of walls	5.2 kN/m²
Ultimate design load on the column	300 kN
Ultimate design uniformly distributed load on the beam	30 kN

Ultimate design load on the wall from above   100 kN
Category of manufacturing control   special
Category of construction control   normal
Effective height of the wall   3.0 m
Solid concrete blocks with a height to least horizontal dimension equal to 1.4 are to be
used.

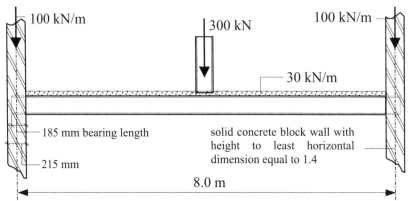

Figure 2.34

Solution: Consider a 1.0 metre length of wall
Characteristic self-weight of wall $= 1.4 \times (0.4 \times 4.25 \times 5.2)$ $=$ 12.4 kN/m
Total design load on the beam $= [300 + (8.0 \times 30)] =$ 540 kN
End reaction on wall due to the beam $= (0.5 \times 540)$ $=$ 270 kN/wall
Load on wall from above (considered concentric) $=$ 100 kN/m
Total design load on the wall/m length $= (270 + 100 + 12.4) =$ **382.4 kN**

Clause 32.2.1 *Design Vertical Load Resistance of Walls*

Design vertical load resistance/unit length $= \dfrac{\beta\, t\, f_k}{\gamma_m}$

Clause 27.3   *Partial Safety Factor for Material Strength (γ_m)*
Category for manufacturing control is special
Category for construction control is normal

Table 4

Table 4. Partial safety factors for material strength, γ_m		Category of construction control	
		Special	**Normal**
Category of manufacturing control of structural units	**Special**	2.5	3.1
	Normal	2.8	3.5

Partial safety factor $\gamma_m = $ **3.1**

Clause 28   *Consideration of Slenderness of Walls and Columns*
                    slenderness ratio   SR $=$ h_{ef}/t_{ef} $\leq$ 27

Clause 28.3.1 *Effective Height*
                    h_{ef} =   3000 mm

Clause 28.4.1 *Effective Thickness*
For single-leaf walls the effective thickness is equal to the actual thickness as indicated in Figure 3 of the code.

$$t_{ef} = 215 \text{ mm}$$

$$SR = \frac{3000}{215} = \mathbf{13.9} < 27$$

Clause 31   *Eccentricity Perpendicular to the Wall*
The design loading on the wall comprises a concentric element from above and an eccentric element from the beam reaction. These two loads must be considered together in an equivalent system as shown in Figure 2.35, to determine the eccentricity due to both acting simultaneously.

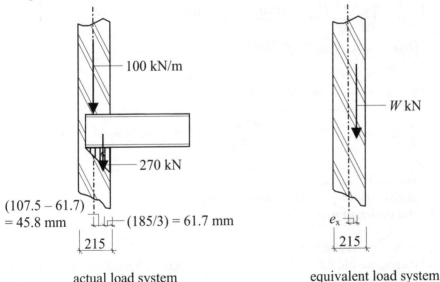

actual load system   equivalent load system

Figure 2.35

The equivalent load system must have the same vertical load and the same bending moment about the centre-line as the actual system.

Equating loads W $=$ $(100 + 270) =$ 370 kN
Equating moments $(W \times e_x)$ $=$ $(270 \times 45.8) =$ 12366 kNmm
            $\therefore$ e_x $=$ $(12366/370) =$ 33.4 mm

In Table 7 the eccentricity is expressed in terms of the thickness t,

$$e_x = \frac{33.4}{215}t = 0.16t$$

Table 7   Capacity Reduction Factor
Linear interpolation between slenderness and eccentricity values is permitted when using Table 7.

Table 7. Capacity reduction factor, β				
Slender-ness ratio h_{ef}/t_{ef}	Eccentricity at top of wall, e_x			
	Up to 0.05t (see note 1)	0.1t	0.2t	0.3t
0	1.00	0.88	0.66	0.44
12	0.93	0.87	0.66	0.44
14	0.89	0.83	0.66	0.44
16	0.83	0.77	0.64	0.44
18	0.77	0.70	0.57	0.44

$$\beta_{(SR = 13.9;\ ex = 0.16t)} = 0.73$$

Clause 32.2.1   Design Vertical Load Resistance $= \dfrac{\beta\, t\, f_k}{\gamma_m}$

$$f_k \geq \frac{392.4 \times \gamma_m}{\beta\, t} = \frac{382.4 \times 3.1}{0.73 \times 215} = \textbf{7.55 N/mm}^2$$

Table 1   Requirements for Mortar
          Required masonry cement : sand ratio is 1 : 3
          Mortar designation (ii) is suitable

Clause 23.1.6   Solid Concrete Block Walls
When using solid concrete blocks with $0.6 < H:L < 2.0$ interpolate between the values given in Tables 2(b) 2(d).

Table 2(d)   Characteristic Compressive Strength of Masonry
Assuming a unit of strength equal 15.0 N/mm^2 combined with mortar type (ii) will satisfy the masonry strength requirement (i.e. f_k = 8.3 N/mm^2 > 7.55 N/mm^2).

Note: The design loads calculated above include the self-weight of a height of wall equal to 0.4h since the critical section for maximum eccentricity occurs at this level as indicated in Section 2.1.7. The self-weight of the full height of the wall is often used instead of 0.4h, this slightly overestimates the design axial load and is conservative.

2.4   Example 2.3   Single-Leaf Masonry Wall 3

An internal brick wall supports a two-span reinforced concrete floor slab as shown in Figure 2.36. Using the design data given, determine a suitable brick/mortar combination.

Design data:

Self-weight of 102.5 mm thick standard format brickwork	1.8 kN/m^2
Self-weight of plaster 12 mm thick	0.3 kN/m^2
Characteristic dead load on floor slab	5.0 kN/m^2
Characteristic imposed load on floor slab	1.5 kN/m^2
Characteristic dead load from wall above	120.0 kN/metre length
Characteristic imposed load from wall above	20.0 kN/metre length
Category of manufacturing control	normal
Category of construction control	normal

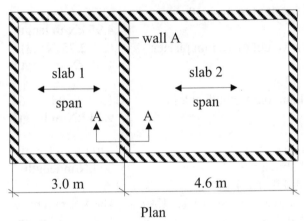

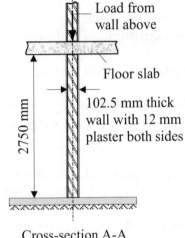

Plan   Cross-section A-A

Figure 2.36

Solution: Consider a 1.0 metre length of wall

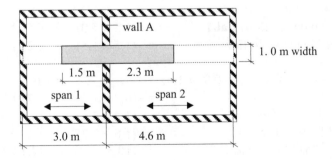

The wall supports an area equal to (1.5 × 1.0) from span 1 and an area equal to (2.3 × 1.0) from span 2.

The loads due to the self-weight of the wall and from the wall above are considered to be concentric whilst those from the floor slab are considered to be eccentric as indicated in Clause 31 of the code, i.e.

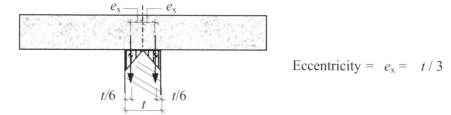

Eccentricity = e_x = $t/3$

Each side of the floor may be taken as being supported individually on half the total bearing area.

Concentric Loads:

Characteristic dead load due to the self-weight of the wall		= (2.75×1.8)
		= **4.95 kN/m length**
Characteristic dead load due to the self-weight of 12 mm plaster		= $(2 \times 2.75 \times 0.3)$
		= 1.65 kN/m length

Clause 22 Load Factor $\gamma_{f\ dead\ load}$ = 1.4
Design load due to self-weight of wall and plaster both sides   = $1.4 \times (4.95 + 1.65)$
                                                                    = **9.24 kN/m length**

Characteristic dead load from wall above		= 120 kN/m length
Characteristic imposed load from wall above		= 20 kN/m length

Clause 22 Load Factors $\gamma_{f\ dead\ load}$ = 0.9 or 1.4; $\gamma_{f\ imposed\ load}$ = 1.6

Minimum design dead load from wall above	= (0.9×120)	= **108 kN/m length**
Maximum design dead load from wall above	= (1.4×120)	= 168 kN/m length
Design imposed load from wall above	= (1.6×20)	= 32 kN/m length
Maximum total design load from wall above	= $(168 + 32)$	= **200 kN/m length**

Eccentric Loads:

Characteristic dead load due to floor **slab1**	= (1.5×5.0)	= 7.5 kN/m length
Characteristic imposed load due to floor slab1	= (1.5×1.5)	= 2.25 kN/m length

Clause 22 Load Factors $\gamma_{f\ dead\ load}$ = 0.9 or 1.4; $\gamma_{f\ imposed\ load}$ = 1.6

Minimum design dead load from slab 1	= (0.9×7.5)	= **6.75 kN/m length**
Maximum design dead load from slab1	= (1.4×7.5)	= 10.5 kN/m length
Design imposed load from slab 1	= (1.6×2.25)	= 3.6 kN/m length
Maximum total design load from slab 1	= $(10.5 + 3.6)$	= **14.1 kN/m length**

Characteristic dead load due to floor **slab2**	= (2.3×5.0)	= 11.5 kN/m length
Characteristic imposed load due to floor slab2	= (2.3×1.5)	= 3.45 kN/m length

Clause 22 Load Factors $\quad\quad \gamma_{\text{f dead load}} = 1.4; \quad\quad \gamma_{\text{f imposed load}} = 1.6$
Maximum total design load from slab 2 $\quad = \quad [(1.4 \times 11.5) + (1.6 \times 3.45)]$
$$= \textbf{21.62 kN/m length}$$

There are three load combinations which should be considered:

Case 1: *Maximum vertical loads*

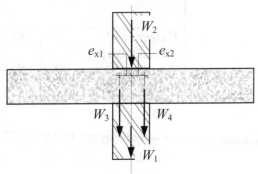

$\quad W_1 \quad = \quad 9.24 \text{ kN/m length}$
$\quad W_2 \quad = \quad 200 \text{ kN/m length}$
$\quad W_3 \quad = \quad 14.1 \text{ kN/m length}$
$\quad W_4 \quad = \quad 21.62 \text{ kN/m length}$
$\quad \textbf{Total} = \textbf{244.96 kN/m length}$

Case 2: *Maximum moment loads*

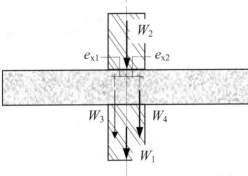

$\quad W_1 \quad = \quad 9.24 \text{ kN/m length}$
$\quad W_2 \quad = \quad 200 \text{ kN/m length}$
$\quad W_3 \quad = \quad 6.75 \text{ kN/m length}$
$\quad W_4 \quad = \quad 21.62 \text{ kN/m length}$
$\quad \textbf{Total} = \textbf{237.61 kN/m length}$

Case 3: *Maximum eccentricity loads*

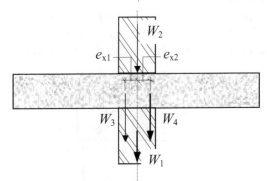

$\quad W_1 \quad = \quad 9.24 \text{ kN/m length}$
$\quad W_2 \quad = \quad 108 \text{ kN/m length}$
$\quad W_3 \quad = \quad 6.75 \text{ kN/m length}$
$\quad W_4 \quad = \quad 21.62 \text{ kN/m length}$
$\quad \textbf{Total} = \textbf{145.61 kN/m length}$

Clause 32.2.1 Design Vertical Load Resistance of Walls

Design vertical load resistance/unit length $\quad = \quad \dfrac{\beta\, t\, f_k}{\gamma_m}$

Clause 27.3 Partial Safety Factor for Material Strength (γ_m)
Category for manufacturing control is normal
Category for construction control is normal

Table 4

Table 4. Partial safety factors for material strength, γ_m		Category of construction control	
		Special	Normal
Category of manufacturing control of structural units	Special	2.5	3.1
	Normal	2.8	3.5

γ_m = **3.5**

Clause 28   *Consideration of Slenderness of Walls and Columns*
                 slenderness ratio   SR   = h_{ef}/t_{ef} ≤ 27

Clause 28.3.1   *Effective Height*
                  h_{ef} = (0.75×2750) = 2062.5 mm

Clause 28.4.1   *Effective Thickness*
For single-leaf walls the effective thickness is equal to the actual thickness as indicated in Figure 3 of the code.
$$t_{ef} = 102.5 \text{ mm}$$
$$SR = \frac{2062.5}{102.5} = \mathbf{18.78} < 27$$

CASE 1:
Clause 31   *Eccentricity Perpendicular to the Wall*
The design loading on the wall comprises a concentric element from above and two eccentric elements from the slab reactions.

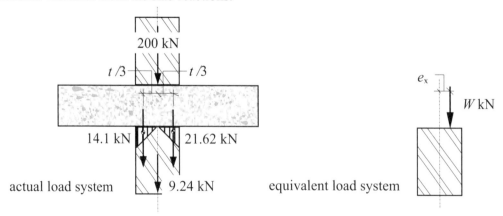

The equivalent load system must have the same vertical load and the same bending moment about the centre-line as the actual system.
Equating vertical forces W = $(200 + 14.1 + 21.62)$ = 235.7 kN
Equating moments $(235.7 \times e_x)$ = $[(21.62 \times t/3) - (14.1 \times t/3)]$ = 2.51t kNmm

$$\therefore \quad e_x \; = \; (2.51t/235.7) \quad = \; 0.01t \text{ mm}$$
$$< \; 0.05t$$

Maximum vertical load $\;=\; 244.96$ kN

Note: The self-weight of the wall and plaster has not been included in the calculation for the eccentricity.

Table 7 Capacity Reduction Factor
Linear interpolation between slenderness and eccentricity values is permitted when using Table 7.

Table 7. Capacity reduction factor, β				
Slender-ness ratio h_{ef}/t_{ef}	**Eccentricity at top of wall,** e_x			
	Up to 0.05t (see note 1)	0.1t	0.2t	0.3t
0	1.00	0.88	0.66	0.44
18	0.77	0.70	0.57	0.44
20	0.70	0.64	0.51	0.37
22	0.62	0.56	0.43	0.30

$$\text{SR} = 20.12 \qquad \beta_{ex=0.01} \; = \; [0.70 - (0.08 \times 0.12/2.0)] = \; 0.7$$

Clause 32.2.1   Design Vertical Load Resistance $\quad = \; \dfrac{\beta\, t\, f_k}{\gamma_m} \; \geq \; 244.96$ kN

$$f_k \; \geq \; \frac{244.96 \times \gamma_m}{\beta\, t} \; = \; \frac{244.96 \times 3.5}{0.7 \times 102.5} \; = \; \textbf{11.9 N/mm}^2$$

CASE 2:
Clause 31   Eccentricity Perpendicular to the wall

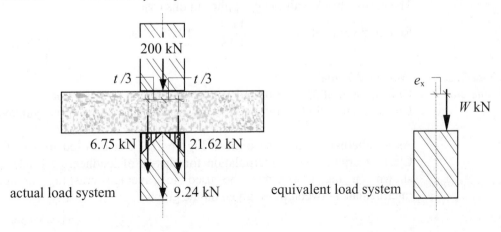

200 kN

$t/3$ $t/3$

e_x

W kN

6.75 kN   21.62 kN

actual load system   9.24 kN   equivalent load system

Equating vertical forces W = $(200 + 6.75 + 21.62)$ =   228.37 kN
Equating moments $(228.37 \times e_x)$ = $[(21.62 \times t/3) - (6.75 \times t/3)]$ = $4.96t$ kNmm
                   $\therefore$ e_x = $(4.96t/228.37)$ = $0.02t$ mm
                                                                < $0.05t$

         Maximum vertical load   =   237.61 kN
         Since e_x is also $< 0.05t$ Case (1) is more severe.

CASE 3:
Clause 31   *Eccentricity Perpendicular to the wall*

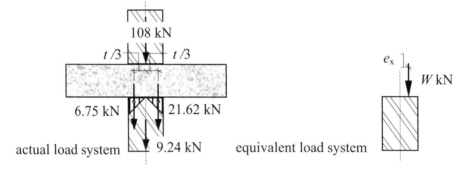

Equating vertical forces W = $(108 + 6.75 + 21.62)$ =   136.37 kN
Equating moments $(136.37 \times e_x)$ = $[(21.62 \times t/3) - (6.75 \times t/3)]$ = $4.96t$ kNmm
                   $\therefore$ e_x = $(4.96t/136.37)$ = $0.04t$ mm
                                                                < $0.05t$

         Maximum vertical load   =   145.61 kN
         Since e_x is also $< 0.05t$ Case (1) is more severe.
         Minimum value of f_k $\geq$ 11.9 N/mm^2

Note: In most situations Case (1) will be the critical case, however, it is important for designers to be aware of the other possibilities.

Clause 23.1.2   *Narrow Brick Walls*
                   The narrow brick wall factor applies in this case.
                   Required value of f_k $\geq \dfrac{11.9}{1.15} =$ **10.35 N/mm^2**

Clause 23.1   *Normal Masonry*
Table 2(a)   Unit strength of 35 N/mm^2 and mortar designation (i)   or
                   Unit strength = 50 N/mm^2 and mortar strength (i), (ii) or (iii) are suitable.

                   As an alternative to these options Figure 1(a) can be used to specify a Class of brick. Linear interpolation for classes of loadbearing bricks not shown on the graph may be used for average crushing strengths intermediate between those given on the graph.

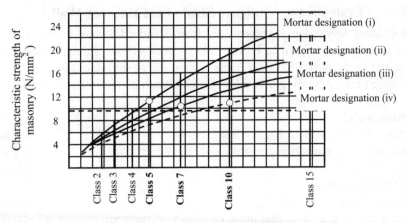

Figure 1(a) from the code   Compressive strength of unit (N/mm²)

2.5   Stiffened Walls

The introduction of piers to stiffen walls reduces their slenderness and increases the loadbearing capacity. The reduction in slenderness results from an effective thickness which is greater than the unstiffened wall and is evaluated using a stiffness coefficient as indicated in Figure 3, Table 5 and Clause 28.4.2 in the code, i.e.

$$t_{ef} = t \times K \quad \text{where} \quad K \geq 1.0$$

The effective thickness is subsequently used to determine the slenderness ratio and hence the capacity reduction factor β. Clearly since K is ≥ 1.0 the slenderness will be less, and consequently the value of β will be greater, than that for the unstiffened wall. In addition to this an equivalent wall thickness can be estimated since the pier will support some of the applied load. This can be carried out in most cases as follows using a 'rule of thumb' as indicated in Figure 2.37.

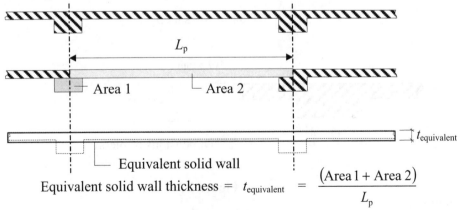

$$\text{Equivalent solid wall thickness} = t_{equivalent} = \frac{(\text{Area 1} + \text{Area 2})}{L_p}$$

Figure 2.37

The value of $t_{equivalent}$ can be used in the equation given in Clause 32.2.1, i.e. $\dfrac{\beta t f_k}{\gamma_m}$.

2.6   Example 2.4   Stiffened Single-Leaf Masonry Wall

An existing workshop is constructed from standard format bricks with a rendered single-skin and piers as shown in Figure 2.38. A series of steel beams supported as indicated are required to support additional loads. Using the design data given determine the maximum ultimate vertical loading/metre length which can be applied to the wall.

Design data:

Category of manufacturing control	normal
Category of construction control	normal
Brick strength	Class 7
Mortar designation	Type (ii)
Effective height of wall (h_{ef})	3200 mm
In this problem assume that the eccentricity of the applied load is	$< 0.05t$

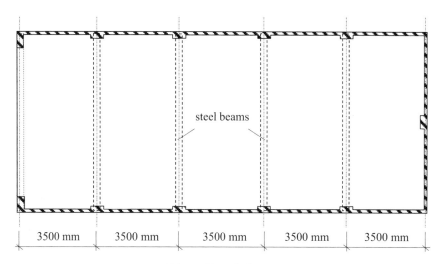

Plan of workshop

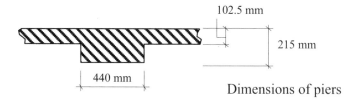

Dimensions of piers

Figure 2.38

Solution:
Clause 32.2.1

$$\text{Design vertical load resistance} = \frac{\beta\, t\, f_k}{\gamma_m}$$

Figure 1(a)   *Characteristic Compressive Strength of Brick Masonry (f_k)*
              Class 7 bricks / Mortar designation (ii) $f_k \approx 11.8 \text{ N/mm}^2$

Table 4
              Manufacturing control – normal / Construction control – normal
              Partial safety factor $\gamma_m = 3.5$

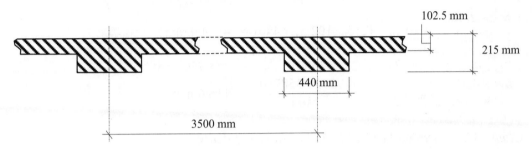

Clause 28.4.2 and Figure 3 Effective Thickness
              $t_{ef} = t \times K$

Table 5   *Stiffness Coefficient for Walls Stiffened by Piers*

Table 5. Stiffness coefficient for walls stiffened by piers

Ratio of pier spacing (centre to centre) to pier width	Ratio t_p/t of pier thickness to actual thickness of wall to which it is bonded		
	1	2	3
6	1.0	1.4	2.0
10	1.0	1.2	1.4
20	1.0	1.0	1.0

NOTE. Linear interpolation between the values given in table 5 is permissible, but not extrapolation outside the limits given.

$$\frac{\text{pier spacing}}{\text{pier width}} = \frac{3500}{440} \approx 8.0; \qquad \frac{\text{pier thickness}}{\text{actual thickness}} = \frac{215}{102.5} = 2.1$$

Using linear interpolation in Table 5 $K = 1.34$
              $t_{ef} = (1.34 \times 102.5) = 137.4 \text{ mm}$

Clause 28   *Slenderness*
              $$SR = \frac{3200}{137.4} = \mathbf{23.3} < 27$$

Table 7   *Capacity Reduction Factor* (β)

Using linear interpolation in Table 7 $\beta = 0.56$

The equivalent thickness to allow for the additional load carried by the piers can be estimated:

$$\text{Equivalent solid wall thickness} = t_{\text{equivalent}} = \frac{(\text{Area 1} + \text{Area 2})}{L_{\text{p}}}$$

$$\text{Area 1} = [(215 - 102.5) \times 440] = 49{,}500 \text{ mm}^2$$

$$\text{Area 2} = (102.5 \times 3500) = 358{,}750 \text{ mm}^2$$

$$t_{\text{equivalent}} = \frac{(49500 + 358750)}{3500} = 116.6 \text{ mm}$$

Clause 32.2.1 Design Vertical Load Resistance of Walls

$$\text{Design vertical load resistance} = \frac{\beta \, t_{\text{equivalent}} \, f_k}{\gamma_{\text{m}}} = \frac{0.56 \times 116.6 \times 11.8}{3.5}$$

$$= \mathbf{220.1 \ kN/m}$$

2.7   Cavity Walls

The fundamental requirements of external walls in buildings include the provision of adequate strength, stability, thermal and sound insulation, fire resistance and resistance to rain penetration. Whilst a single skin wall can often satisfy the strength and stability requirements, in many cases the thickness of wall necessary to satisfy some of the other requirements (e.g. resistance to rain penetration c.f. BS 5628 : Part 3), can be uneconomic and inefficient. In most cases of external wall design, cavity wall construction comprising two leaves with a gap between them is used. Normally the outer leaf is a half-brick (102.5 mm thick) common or facing brick and the inner leaf is either the same or light weight, thermally efficient, concrete block. The minimum thickness of each leaf is specified in Clause 29.1.2 of BS 5628 : Part 1 as 75 mm. The width of the cavity between the leaves may vary between 50 mm and 150 mm but should not be greater than 75 mm where either of the leaves is less than 90 mm in thickness as indicated in Clause 29.1.3 of the code. These criteria are illustrated in Figure 2.39.

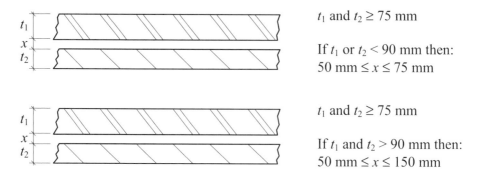

t_1 and $t_2 \geq 75$ mm

If t_1 or $t_2 < 90$ mm then:
50 mm $\leq x \leq 75$ mm

t_1 and $t_2 \geq 75$ mm

If t_1 and $t_2 > 90$ mm then:
50 mm $\leq x \leq 150$ mm

Figure 2.39

The minimum width requirement is to reduce the possibilty of bridging across the cavity, e.g. by mortar droppings, causing the transmission of moisture. The maximum width requirement is to limit the length of the wall ties connecting the two leaves and hence reduce their tendency to buckle if subjected to compressive forces.

The effect of ties is to stiffen each of the leaves in a wall and consequently reduce the slenderness ratio.

Where the applied vertical load is supported by one leaf only, the influence of the unloaded leaf is considered when evaluating the effective thickness of the wall, as indicated in Figure 3 of the code. The cross-sectional area used to determine the load resistance is that of the loaded leaf only.

Where the applied vertical load acts between the centroids of the two leaves it should be replaced by a statically equivalent axial load in each leaf as indicated in Clause 32.2.3 of the code and illustrated in Figure 2.40.

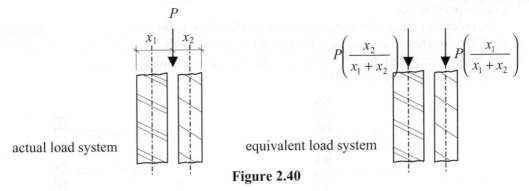

Figure 2.40

Each leaf is designed seperately to resist the equivalent axial load using the stiffening effect of the other leaf to determine the effective thickness and hence the slenderness ratio.

The provision of ties is governed by the requirements of Clause 29.1.5 and suitable minimum values are given Table 6 of BS 5628 : Part 1 and also shown in Figure 2.41.

Selection of wall ties: types and lengths				
Least leaf thickness (one or both)	Nominal cavity width	Permissible type of tie		Tie length
		Shape name* in accordance with BS 1243	Type member** in accordance with DD 140 : Part 2	
mm	mm			mm
75	75 or less	(a), (b) or (c)	1,2,3,4	175
90	75 or less	(a), (b) or (c)	1,2,3,4	200
90	76 to 90	(b) or (c)	1 or 2	225
90	91 to 100	(b) or (c)	1 or 2	225
90	101 to 125	(c)	1 or 2	250
90	126 to 150	(c)	1 or 2	275
* (a) – butterfly (b) – double triangle (c) – vertical twist				
** type 1 is the stiffest and type 4 the least stiff				

Figure 2.41

The ties should be staggered and evenly distributed over a wall area. Additional ties should be provided at a rate of one tie per 300 mm height or equivalent, located not more than 225 mm from the vertical edges of openings and at vertical unreturned or unbonded edges such as at movement joints or the sloping verge of gable walls. As indicated in Clause 29.1.6, ties should be embedded at least 50 mm in each leaf.

Since the two leaves of a cavity wall may be of different materials with differing physical properties and/or subject to different thermal effects differential movement is inevitable. To prevent potentially damaging loosening of the embedded ties the uninterrupted height of the outer leaf of external cavity walls can be limited or suitable detailing of the construction incorporated to accommodate the movement.

2.7.1   *Limitation on Uninterrupted Height   (Clause 29.2.2)*

The code specifies that:

'…. the outer leaf should be supported at intervals of not more than every third storey or every 9 m, whichever is less. …..for buildings not exceeding four storeys or 12 m in height, whichever is less, the outer leaf may be uninterrupted for its full height.'

This is illustrated in Figure 2.42.

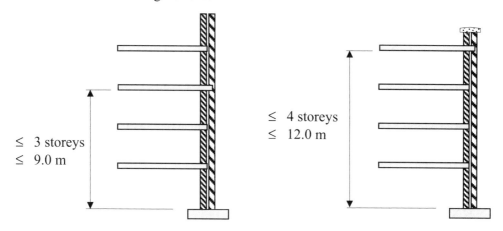

$\leq$ 3 storeys
$\leq$ 9.0 m

$\leq$ 4 storeys
$\leq$ 12.0 m

Figure 2.42

2.7.2   *Accommodation of Differential Vertical Movement (Clause 29.2.3)*

Where differential movement is accommodated by calculation this should include all factors such as elastic movement, moisture movement, thermal movement, creep etc., and the value should not exceed 30 mm. Detailing which includes separate lintols for both leaves, appropriate ties which can accommodate movement and the fixing of window frames cills etc. must be considered. In most cases engineers adopt the method of limiting the uninterrupted height.

2.7.3   *Accommodation of Differential Horizontal Movement (Clause 29.2.4)*

Advice is given in BS 5628 : Part 3 on the provision of vertical joints to accommodate horizontal movements.

2.8   Example 2.5   Cavity Wall 1

Consider the structure shown in Figure 2.43 in which the external walls supporting the roof slab are cavity walls comprising:

i)   both leaves of standard format bricks and a cavity width of 50 mm,
ii)   a standard format brick outer leaf, a 100 mm hollow block (with a height-to-width ratio of 0.6), inner leaf and a cavity width of 75 mm,
iii)   a standard format brick outer leaf, a 90 mm × 90 mm modular brick inner leaf and a cavity width of 50 mm.

Using the design data given determine the maximum ultimate vertical load which can be supported by the inner leaf.

Design data:
Category of manufacturing control   special
Category of construction control   special
Mortar designation/Compressive strength of unit   case (i)   Type (i) / 20 N/mm^2
                                                                                             case (ii)   Type (ii) / 10 N/mm^2
                                                                                             case (iii) Type (iii) / 15 N/mm^2

Assume that the walls are part of a braced structure.

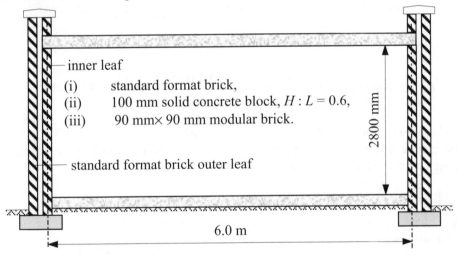

Figure 2.43

Solution: Consider a 1.0 metre length of wall

(i) Standard format bricks

Clause 32.2.1   Design Vertical Load Resistance of Walls

$$\text{Design vertical load resistance/unit length} = \frac{\beta\, t\, f_k}{\gamma_m}$$

Clause 27.3   *Partial Safety Factor for Material Strength (γ_m)*
Category for manufacturing control is special
Category for construction control is special
Table 4 γ_m = **2.5**

Clause 28   *Consideration of Slenderness of Walls and Columns*
slenderness ratio (SR) = h_{ef}/t_{ef} ≤ 27

Clause 28.2.2 Horizontal Lateral Support
Since this structure has a concrete roof with a bearing length of at least one-half the thickness of the wall, enhanced resistance to lateral movement can be assumed.
Clause 28.3.1 Effective Height
h_{ef} = 0.75 × clear distance between lateral supports
     = (0.75 × 2800)   = 2100 mm

Clause 28.4.1 Effective Thickness
For cavity walls the effective thickness is as indicated in Figure 3 of the code and equal to the greatest of:

(a) $2/3(t_1 + t_2)$ = $2/3(102.5 + 102.5)$ = 136.7 mm   or
(b) t_1 = 102.5 mm   or
(c) t_2 = 102.5 mm

∴ t_{ef} = 136.7 mm

$$\text{Slenderness ratio} = \text{SR} = \frac{2100}{136.7} = 15.4 < 27$$

Clause 31   *Eccentricity Perpendicular to the wall*
The load maybe assumed to act at an eccentricity equal to one-third of the depth of the bearing area from the loaded face of the wall.

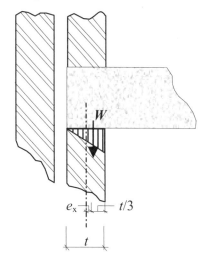

eccentricity (e_x) = $(1/2t - 1/3t)$ = $t/6$
                e_x = **0.167t**

Table 7   Capacity Reduction Factor
Linear interpolation between slenderness and eccentricity values is permitted when using Table 7.
β = 0.68

Table 2(a)
Compressive strength of unit   = 20 N/mm^2
Mortar designation   = Type (i)
Compressive strength of masonry
                    f_k = 7.4 N/mm^2

e_x $t/3$

t

Clause 23.1.2   Narrow Brick Walls
When using standard format bricks to construct a wall one brick (i.e. 102.5 mm) wide the values of f_k obtained from Table 2(a) can be multiplied by 1.15. This does **not** apply to the loaded inner leaf of a cavity wall when **both** leaves are loaded (see the '*Handbook to   BS 5628 : Structural Use of Masonry*', ref. 36).

$$f_k = (7.4 \times 1.15) = \mathbf{8.51 \ N/mm^2}$$

Clause 32.2.1   Design Vertical Load Resistance $\quad = \dfrac{\beta t f_k}{\gamma_m} = \dfrac{0.68 \times 102.5 \times 8.51}{2.5}$

$$= \mathbf{273.3 \ kN/m}$$

(ii)   100 mm hollow blocks with a height-to-width ratio of 0.6

Clause 27.3   Partial Safety Factor for Material Strength (γ_m)
                 Category for manufacturing control is special
                 Category for construction control is special
Table 4 $\gamma_m = \mathbf{2.5}$

Clause 28   *Consideration of Slenderness of Walls and Columns*
                 slenderness ratio (SR) $= h_{ef}/t_{ef} \quad \leq \ 27$

Clause 28.2.2   Horizontal Lateral Support
Since this structure has a concrete roof with a bearing length of at least one-half the thickness of the wall, enhanced resistance to lateral movement can be assumed.

Clause 28.3.1   Effective Height
                 $h_{ef} = \ 0.75 \times$ clear distance between lateral supports
                 $= (0.75 \times 2800) \ = \ 2100$ mm

Clause 28.4.1   Effective Thickness
For cavity walls the effective thickness is as indicated in Figure 3 of the code and equal to the greatest of:

       (a) $2/3(t_1 + t_2)$ $\quad = \ 2/3(100 + 102.5) \quad = \ 135$ mm $\qquad$ or
       (b) $\quad t_1$ $\qquad\quad = \ 100$ mm $\qquad$ or
       (c) $\quad t_2$ $\qquad\quad = \ 102.5$ mm

$$\therefore \ t_{ef} = \ 135 \ \text{mm}$$

Slenderness ratio $\ = \ $ SR $\ = \ \dfrac{2100}{135} = \ 15.6 \ < 27$

Clause 31   Eccentricity Perpendicular to the Wall
The load maybe assumed to act at an eccentricity equal to one-third of the depth of the bearing area from the loaded face of the wall.

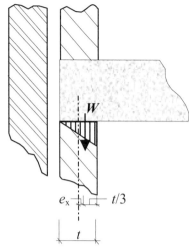

eccentricity $(e_x) = (1/2t - 1/3t) = t/6$

$$e_x = \mathbf{0.167}t$$

Table 7   Capacity Reduction Factor
Linear interpolation between slenderness and eccentricity values is permitted when using Table 7.

$\beta = 0.69$

Table 2(b)
Compressive strength of unit   = 10 N/mm²
Mortar designation   = Type (ii)
Compressive strength of masonry

$$f_k = 4.2 \text{ N/mm}^2$$

Clause 32.2.1   Design Vertical Load Resistance $= \dfrac{\beta\, t\, f_k}{\gamma_m} = \dfrac{0.69 \times 100 \times 4.2}{2.5}$

$$= \mathbf{119.3 \text{ kN/m}}$$

(**Note:** The narrow brick wall factor applies only to standard format bricks.)

(iii)   90 mm × 90 mm modular bricks
Note: It is not unusual to include a course or two of bricks or cut block in an inner block-leaf wall to accommodate coursing with a standard brick outer-leaf.

Clause 27.3   Partial Safety Factor for Material Strength (γ_m)
            Category for manufacturing control is special
            Category for construction control is special
Table 4 $\gamma_m = \mathbf{2.5}$

Clause 28   Consideration of Slenderness of Walls and Columns
            slenderness ratio (SR) = h_{ef}/t_{ef} $\leq$ 27

Clause 28.2.2   Horizontal Lateral Support
Since this structure has a concrete roof with a bearing length of at least one-half the thickness of the wall, enhanced resistance to lateral movement can be assumed.

Clause 28.3.1   Effective Height
            h_{ef} = 0.75 × clear distance between lateral supports
                = (0.75 × 2800) = 2100 mm

Clause 28.4.1   Effective Thickness
For cavity walls the effective thickness is as indicated in Figure 3 of the code and equal to the greatest of:

(a) $2/3(t_1 + t_2)$ = 2/3(90 + 102.5)   = 128.3 mm   or
(b) t_1 = 90 mm   or
(c) t_2 = 102.5 mm
            $\therefore t_{ef}$ = 128.3 mm

$$\text{Slenderness ratio} = \text{SR} = \frac{2100}{128.3} = 16.4 < 27$$

Clause 31 Eccentricity Perpendicular to the Wall
The load maybe assumed to act at an eccentricity equal to one-third of the depth of the bearing area from the loaded face of the wall.

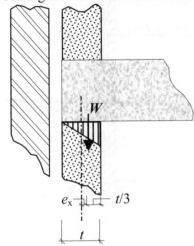

eccentricity $(e_x) = (1/2t - 1/3t) = t/6$
$$e_x = 0.167t$$

Table 7 Capacity Reduction Factor
Linear interpolation between slenderness and eccentricity values is permitted when using Table 7.
$$\beta = 0.67$$

Clause 23.1.3 and Table 2(a)
Compressive strength of unit $= 15 \text{ N/mm}^2$
Mortar designation $= $ Type (iii)
Compressive strength of masonry:
$$f_k = (1.25^* \times 5.0) \qquad = 6.25 \text{ N/mm}^2$$

* **Note:** A narrow brick wall factor is incorporated in the 1.25 value given when using modular bricks as indicated in this Clause 23.1.3.

Clause 32.2.1 Design Vertical Load Resistance $= \dfrac{\beta\, t\, f_k}{\gamma_m} = \dfrac{0.67 \times 100 \times 6.25}{2.5}$
$$= \mathbf{167.5 \text{ kN/m}}$$

2.9 Example 2.6 Cavity Wall 2

Consider the structure shown in Figure 2.44 in which the external walls supporting the roof slab are cavity walls comprising leaves of standard format bricks and a cavity width of 50 mm. Using the design data given check the suitability of the inner leaf of the wall to support the loads given.

Design data:

Category of manufacturing control	normal
Category of construction control	normal
Mortar designation / Compressive strength of unit	Type (ii) / 15 N/mm^2
Characteristic dead load on roof (including self-weight)	4.5 kN/m^2
Characteristic imposed load on roof	1.5 kN/m^2
Characteristic dead load on all floors (including self-weight)	5.0 kN/m^2
Characteristic imposed load on all floors	2.5 kN/m^2
Characteristic self-weight of walls	2.0 kN/m^2

Note: assume that the walls are part of a braced structure and that lateral forces due to wind loading are not being considered in this example.

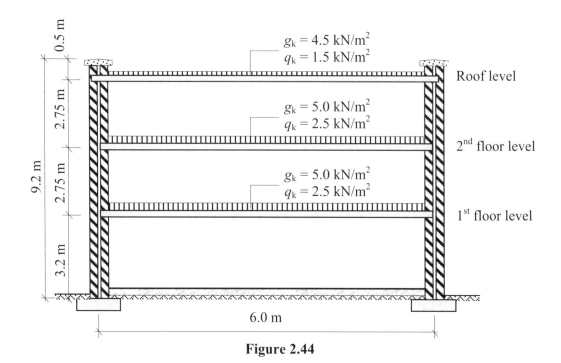

Figure 2.44

Solution:
Consider a 1.0 metre length of wall and check the capacity of the wall at the underside of the first floor level.

Loading:

Characteristic dead load due to roof slab	$= (4.5 \times 3.0)$	$= 13.5$ kN/m length
Characteristic imposed load due to roof slab	$= (1.5 \times 3.0)$	$= 4.5$ kN/m length
Characteristic dead load due to floor slabs	$= [2 \times (5.0 \times 3.0)]$	$= 30.0$ kN/m length
Characteristic imposed load due to floor slabs	$= [2 \times (2.5 \times 3.0)]$	$= 15.0$ kN/m length

In BS 6399 : Part 1 : 1996 (Loading for Buildings), an allowance is made for the imposed loading to be considered when designing multi-storey columns, walls, their supports and foundations. In Table 2 of the code a reduction of 10% in the total distributed imposed load on all floors carried by the member under consideration is permitted when that member supports two floors.

Note: The % reduction *does not* apply to roof loading.

Ultimate design load due to roof slab	$=$	$[(1.4 \times 13.5) + (1.6 \times 4.5)]$
	$=$	26.1 kN/m length
Ultimate design load due to floor slabs	$=$	$[(1.4 \times 30) + (1.6 \times 15.0)]$
	$=$	66.0 kN/m length
10% reduction of imposed floor loads	$=$	$- [0.1 \times (1.6 \times 15.0)]$
	$=$	$-$ 2.4 kN/m
Total design load due to roof and floor slabs	$=$	$(26.1 + 66.0 - 2.4) =$ 89.7 kN/m

Ultimate design load due to self-weight of wall $= (1.4 \times 2.0 \times 9.2 \times 1.0)]$
$= 25.76$ kN/m

Design vertical load applied to inner leaf of wall $= (89.7 + 25.76) = \mathbf{115.5 \ kN/m}$

Clause 32.2.1 Design Vertical Load Resistance of Walls

Design vertical load resistance/unit length $= \dfrac{\beta \, t \, f_k}{\gamma_m}$

Clause 27.3 Partial Safety Factor for Material Strength (γ_m)
Category for manufacturing control is normal
Category for construction control is normal

Table 4 $\gamma_m = \mathbf{3.5}$

Clause 28 Consideration of Slenderness of Walls and Columns
slenderness ratio (SR) $= h_{ef}/t_{ef} \ \leq \ 27$

Note: In buildings of more than two storeys with walls less than 90 mm thick the slenderness should not exceed 20.

Clause 28.2.2 Horizontal Lateral Support
Since this structure has a concrete floor with a bearing length of at least one-half the thickness of the wall, enhanced resistance to lateral movement can be assumed.

Clause 28.3.1 Effective Height
$h_{ef} = \ 0.75 \times$ clear distance between lateral supports
$= \ (0.75 \times 3200) \ = \ 2400$ mm

Clause 28.4.1 Effective Thickness
For cavity walls the effective thickness is as indicated in Figure 3 of the code and equal to the greatest of:

(a) $2/3(t_1 + t_2) \ = \ 2/3(102.5 + 102.5) = \ 136.7$ mm   or
(b) t_1 $= \ 102.5$ mm   or
(c) t_2 $= \ 102.5$ mm
$\therefore \ t_{ef} = \ 136.7$ mm

Slenderness ratio $= \ $ SR $\ = \ \dfrac{2400}{136.7} \ = \ 17.6 \ < \ 27$

Clause 31 Eccentricity Perpendicular to the Wall
The load from the first floor slab maybe assumed to act at an eccentricity equal to one-third of the depth of the bearing area from the loaded face of the wall. The loading from the second floor and roof slabs can be considered to be concentric at this level.

The equivalent load system must have the same vertical load and the same bending moment about the centre-line as the actual system.

The critical section normally occurs at 0.4*h* below the level of the 1st. floor slab, see Figure 2.28.

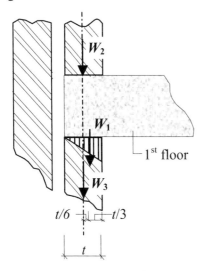

$W_1 = [1.4 \times 15.0 + 1.6 \times 7.5)] = 33.0$ kN/m

Moment induced by the eccentric load is given by

$(W_1 \times t/6.0) = (33.0 \times t)/6.0 = 5.5t$

W_2 is due to a height of brickwork equal to 6.0 m
W_3 is due to a height of brickwork equal to 0.4*h* m

Self-weight of wall considered to be concentric =

$(W_3 + W_2) = [1.4 \times 2.0 \times (6.0 + 1.28)]$
$= 20.38$ kN/m
$e_x = (5.5\,t\,/20.38) = 0.27t$

Table 7   Capacity Reduction Factor
Linear interpolation between slenderness and eccentricity values is permitted when using Table 7.

Table 7. Capacity reduction factor, β				
Slender-ness ratio h_{ef}/t_{ef}	**Eccentricity at top of wall, e_x**			
	Up to 0.05*t* (see note 1)	0.1*t*	0.2*t*	0.3*t*
0	1.00	0.88	0.66	0.44
12	0.93	0.87	0.66	0.44
14	0.89	0.83	0.66	0.44
16	0.83	0.77	0.64	0.44
18	0.77	0.70	0.57	0.44

Table 7   Capacity Reduction Factor
Linear interpolation between slenderness and eccentricity values is permitted when using Table 7.

$$\beta_{(SR = 17.6;\ ex = 0.25t)} = 0.48$$

Table 2(a)

Compressive strength of unit $= 15$ N/mm^2
Mortar designation $=$ Type (ii)
Compressive strength of masonry f_k $= 5.3$ N/mm^2

Clause 23.1.2   Narrow Brick Walls
When using standard format bricks to construct a wall one brick (i.e. 102.5 mm) wide, the values of f_k obtained from Table 2(a) can be multiplied by 1.15.

$$f_k = (5.3 \times 1.15) = 6.1 \text{ N/mm}^2$$

Clause 32.2.1   Design Vertical Load Resistance $= \dfrac{\beta\, t\, f_k}{\gamma_m} = \dfrac{0.48 \times 102.5 \times 6.1}{3.5}$

$$= \textbf{85.7 kN/m} < 115.5 \text{ kN/m}$$

Since the vertical load resistance is less than the applied design load this wall is inadequate. A higher strength unit/mortar combination is required.

2.10   Collar Jointed Walls (Clause 29.6)

Where it is desirable to achieve fairfaced brickwork (i.e. brickwork built with particular care where the finished work is to be visible) on both sides of a 215 mm thick solid wall, for example for architectural reasons, it is necessary to build two leaves of brickwork back-to-back. Both leaves are built in stretcher bond and fairfaced on their front surfaces with a vertical **collar joint** between them as shown in Figure 2.45.

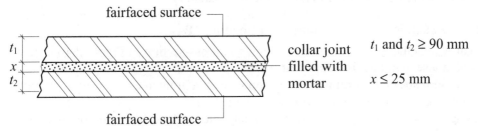

Figure 2.45

Conditions are set out in Clause 29.6 of the code for the design of such walls, they are:

'*Where a wall is constructed of two separate leaves with a vertical collar joint* (i.e. vertical joint parallel to the face of the wall), *not exceeding 25 mm wide then it may be designed as either:*

   (a) a cavity wall, or
   (b) a single leaf wall, provided that the following conditions are satisfied.

      (1) Each leaf is at least 90 mm thick.
      *(2) For concrete brickwork the characteristic compressive strength obtained from Clause **23** is multiplied by 0.9.*
      (3) If the two leaves of the wall are constructed from different materials, it is designed on the assumption that the wall is constructed entirely in the weaker unit. The possibility of differential movement should be taken into account.
      (4) The load is applied to the two leaves and the eccentricity does not exceed

0.2t (except in the case of laterally loaded panels) where t is the overall thickness of the wall.

(1) Flat metal wall ties of cross-sectional area 20 mm × 3 mm at centres not exceeding 450 mm both vertically and horizontally, or an equivalent mesh at the same vertical centres, are provided.

(2) The minimum embedment of the ties into each leaf is 50 mm.

(3) The vertical collar joint between the two leaves is solidly filled with mortar as the work proceeds.'

A collar jointed wall may be designed as a cavity wall or where the conditions *(1)* to *(7)* are satisfied, as a single leaf wall.

2.11   Grouted Cavity Walls (Clause 29.7)

The vertical load capacity of a cavity wall can be increased by the addition of concrete placed in the cavity. In this case the wall may be designed as a single leaf wall with an effective thickness equal to the overall thickness. The code requires that the width of the cavity should be between 50 mm and 100 mm and filled with concrete of a 28-day strength not less than that of the mortar. In addition the requirements detailed in Clause 29.6 (b)(1), (3), (4), (5) and (6), (see section 2.10), should be satisfied.

2.12   Example 2.7   Warehouse and Loading - Bay

A small warehouse and loading bay is shown in cross-section in Figure 2.46. The roof of the warehouse building is supported on the inner leaf of 228 mm cavity walls. The concrete floor slab and loading bay are supported on a grouted cavity wall at the rear and a 327.5 mm thick solid wall at the front as indicated.

Using the design data given determine a suitable brickwork/mortar combination which can be used for:

   (i)   the cavity walls supporting the roof slab,
   (ii)   the grouted cavity wall supporting the floor slab and
   (iii) the solid wall supporting the loading bay.

Design data:

Category of manufacturing control	special
Category of construction control	special
Characteristic dead load on roof (including self-weight)	4.5 kN/m^2
Characteristic imposed load on roof	1.5 kN/m^2
Characteristic dead load on floor slab (including loading bay)	6.0 kN/m^2
Characteristic imposed load on floor slab (including loading bay)	15.0 kN/m^2
Characteristic unit-weight of plaster	21.0 kN/m^3
Characteristic unit-weight of brickwork	18.0 kN/m^3
Characteristic unit-weight of concrete	24.0 kN/m^3

Assume that the walls are part of a braced structure.
Note: Lateral forces due to wind loading are not being considered in this example.

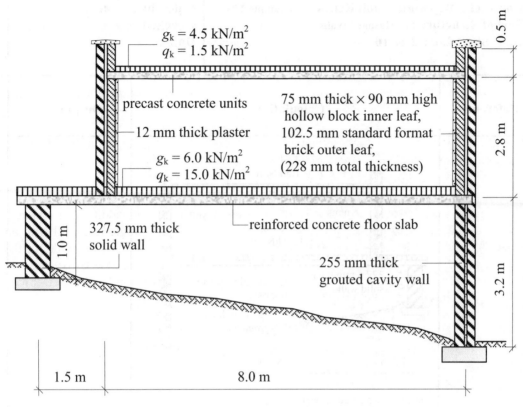

Figure 2.46

The solution for Example 2.7 is presented in a format typical of that used in design office practice when preparing design calculations.

2.12.1 Solution to Example 2.7

Contract : Warehouse Job Ref. No. : Example 2.7 Part of Structure : Masonry Walls Calc. Sheet No. : 1 of 10		Calcs. by : W.McK. Checked by : Date :
References	**Calculations**	**Output**
BS 5628 : Part 1	Structural use of unreinforced masonry Characteristic dead load on roof slab (incl. self wt) = 4.5 kN/m² Characteristic imposed load on roof slab = 1.5 kN/m² Characteristic dead load on floor slab = 6.0 kN/m² Characteristic imposed load on floor slab = 15.0 kN/m²	

References	Calculations	Output

Contract : Warehouse   **Job Ref. No. :** Example 2.7   **Calcs. by :** W.McK.
Part of Structure :   Masonry Walls   **Checked by :**
Calc. Sheet No. : 2 of **10**   **Date :**

Consider a 1m length of wall

Self-weight of 75 mm thick hollow block $\approx$ (0.075×24)
$= 1.8 \text{ kN/m}^2$
Self-weight of 12 mm thick plaster layer $= (0.012 \times 21)$
$= 0.25 \text{ kN/m}^2$
Self-weight of 75 mm thick wall + plaster $= 2.05 \text{ mm}^2$
Self-weight of 255 mm thick grouted cavity wall $\approx (0.255 \times 18)$
$\approx 4.6 \text{ kN/m}^2$

(i) Consider the wall supporting the roof slab:

Clause 22(a)   Design load from roof slab $= (1.4g_k + 1.6q_k)$
$= 4 \times [(1.4 \times 4.5) + (1.6 \times 1.5)]$
$= 34.8 \text{ kN/m length of wall}$

The critical section is at a location $0.4h$ below underside of the roof slab
Design load due to the parapet $= (1.4 \times 1.8 \times 0.5)$
$= 1.26 \text{ kN/m}$

Design load due to the $0.4h$ of blockwork
$= 1.4 \times (0.4 \times 2.8 \times 2.05)$
$= 3.2 \text{ kN/m}$

Design load applied to the inner leaf $= (34.8 + 1.26 + 3.2)$
$= 39.3 \text{ kN/m length}$

Contract : Warehouse Job Ref. No. : Example 2.7	Calcs. by : W.McK.
Part of Structure : Masonry Walls	Checked by :
Calc. Sheet No. : 3 of 10	Date :

References	Calculations	Output
Clause 32.2.1	**Design Vertical Load Resistance of Walls** Design vertical load resistance/unit length $= \dfrac{\beta t f_k}{\gamma_m}$	
Clause 27.3 Table 4	**Partial Safety Factor for Material Strength** (γ_m) Category for manufacturing control is special Category for construction control is special $\gamma_m = 2.5$	
Clause 28	**Consideration of Slenderness of Walls and Columns** slenderness ratio (SR) $= h_{ef}/t_{ef} \quad \le \quad 27$	
Clause 28.2.2	**Horizontal Lateral Support** Since the concrete roof has a bearing length of at least one-half the thickness of the wall, enhanced resistance to lateral movement can be assumed.	
Clause 28.3.1	**Effective Height** $h_{ef} = 0.75 \times$ clear distance between lateral supports $\quad = (0.75 \times 2800) = 2100 \text{ mm}$	
Clause 28.4.1	**Effective Thickness** For cavity walls the effective thickness is as indicated in Figure 3 of the code and equal to the greatest of: (a) $2/3(t_1 + t_2) = 2/3(102.5 + 75) = 118.3 \text{ mm}$ or (b) $\quad t_1 \quad = 102.5 \text{ mm}$ or (c) $\quad t_2 \quad = 75 \text{ mm}$ $\quad \therefore \ t_{ef} = 118.3 \text{ mm}$ Slenderness ratio $= \text{SR} = \dfrac{2100}{118.3} = 17.7 \ < 27$	
Clause 31	**Eccentricity Perpendicular to the Wall** The load from the roof slab maybe assumed to act at an eccentricity equal to one-third of the depth of the bearing area from the loaded face of the wall. The loading from the parapet can be considered to be concentric at this level. The equivalent load system must have the same vertical load and the same bending moment about the centre-line as the actual system. $W_1 = 34.8 \text{ kN}$ Moment induced by the eccentric load is given by: $(W_1 \times t/6.0) = (34.8 \times t)/6.0 = 5.8t$	

References	Calculations	Output

Contract : Warehouse **Job Ref. No. :** Example 2.7
Part of Structure : Masonry Walls
Calc. Sheet No. : 4 of 10

Calcs. by : W.McK.
Checked by :
Date :

$(W_1 + W_2) \times e_x = (5.8 \times t)$
$36.1e_x = 5.8t$

The eccentricity required for the capacity reduction factor β is:

$e_x = (5.8t / 36.1) = 0.16t$

Table 7

Capacity Reduction Factor
Linear interpolation between slenderness and eccentricity values is permitted when using Table 7.

Table 7. Capacity reduction factor, β

Slenderness ratio h_{ef}/t_{ef}	Eccentricity at top of wall, e_x			
	Up to 0.05t (see note 1)	0.1t	0.2t	0.3t
0	1.00	0.88	0.66	0.44
14	0.89	0.83	0.66	0.44
16	0.83	0.77	0.64	0.44
18	0.77	0.70	0.57	0.44

$\beta_{(SR = 17.7;\ ex = 0.16t)} \approx 0.63$

Clause 32.2.1

Design vertical load resistance $= \dfrac{\beta t f_k}{\gamma_m} \geq (36.1 + 3.2)$

$= 39.3$ kN/m length

$f_k = \dfrac{39.3 \times 2.5}{0.63 \times 75} = 2.08$ N/mm^2

Table 2

aspect ratio of concrete block $= \dfrac{height}{least\ horizontal\ dimension}$

$= \dfrac{90}{75} = 1.2$

Clause 23.1.5

For hollow blocks with an aspect ratio between 0.6 and 2.0 f_k should be found by interpolation between the values given in Tables 2(b) and (c).

Contract : Warehouse Job Ref. No. : Example 2.7	Calcs. by : W.McK.
Part of Structure : Masonry Walls	Checked by :
Calc. Sheet No. : 5 of 10	Date :

References	Calculations	Output
Tables 2(b) & (c)	Assume the compressive strength of the unit is 3.5 N/mm² then for any mortar designation (i) to (iv): $f_k = \left[1.7 + (3.5 - 1.7) \times \dfrac{0.6}{1.4}\right]$ = 2.47 N/mm² > 2.08 N/mm² Adopt concrete blocks with unit compressive strength greater than or equal to 3.5 N/mm² and mortar type (i), (ii), (iii) or (iv).	**Adopt concrete blocks with unit strength ≥ 3.5 N/mm²**
Clause 29.7	**(b) Grouted cavity wall supporting the floor slab:** 50 mm ≤ gap between the leaves ≤ 100 mm actual gap = 50 mm and is adequate. **Note:** 28-day strength of concrete infill should not be less than that of the mortar.	
Clause 29.6(b)	(1) each leaf ≥ 90 mm (3) material in each leaf same (4) eccentricity ≤ 0.2*t* To be checked (5) flat metal wall ties of cross-sectional area 20 mm × 3 mm at centres not exceeding 450 mm should be provided in both the vertical and horizontal dimensions (6) the minimum embedment of the ties into each leaf should be at least 50 mm Provided that these conditions are satisfied the grouted wall can be designed as a single wall using an effective thickness equal to the actual overall thickness.	**Ties to satisfy the requirements of Clause 29.6(b)(6)**

References	Calculations	Output

	W_1 is the load due to the inner leaf of the cavity wall above the floor slab. This load is applied at an eccentricity e_1 to the centre-line of the grouted cavity wall given by:
	$e_1 = (100.5 - 37.5) = 63$ mm
	
	Design load from the roof slab $= 34.8$ kN
	Design load from the parapet $= 1.26$ kN
	Design load from the inner leaf $= (1.4 \times 2.8 \times 2.05)$
	                                    $= 8.04$ kN
	Total design load from the inner leaf $= (34.8 + 1.26 + 8.04)$
	                                    $= 44.1$ kN/m length
	
	W_2 is the load due to the outer leaf of the cavity wall above the floor slab. This load is applied at an eccentricity e_2 to the centre-line of the grouted cavity wall given by:
	$e_2 = [127.5 - (0.5 \times 102.5)] = 76.3$ mm
	Design load from the outer leaf $= (1.4 \times 3.2 \times 1.85)$
	    (102.5 thick brickwork) $= 8.3$ kN/m length
	
	W_3 is the load due to $0.4h$ of the grouted cavity wall below the floor slab. This load is applied concentric to the centre-line of the grouted cavity wall and $e_3 = $ zero
	Design load from the grouted cavity wall:
	      $= [1.4 \times (0.4 \times 3.2 \times 4.6)] = 8.2$ kN/m length
	
	W_4 is the load due to the floor slab resting on the grouted cavity wall. This load is assumed to act at one-third of the bearing length from the load-bearing face of the wall at an eccentricity e_4 to the centre-line of the grouted cavity wall given by:
	    $e_4 = [127.5 - (0.333 \times 235)] = 49.2$ mm
	
	Design distributed load on the floor slab is given by:
	    $8 \times [(1.4 \times 6.0) + (1.6 \times 15.0)] = 259.2$ kN (total)
	
	Design point load on the floor slab due to cavity wall is given by:
	    $(W_1 + W_2) = (44.1 + 8.3) = 52.4$ kN	

Clause 31		

Contract : Warehouse Job Ref. No. : Example 2.7	Calcs. by : W.McK.
Part of Structure : Masonry Walls	Checked by :
Calc. Sheet No. : 7 of 10	Date :

References	Calculations	Output
	R_B = [(1.5 × 52.4) + (259.2 × 4.75)]/9.5 = 137.9 kN	

R_B = [(1.5 × 52.4) + (259.2 × 4.75)]/9.5 = 137.9 kN

This reaction is applied at the eccentricity e_4 from the centre-line of the wall.

Total vertical load = $(W_1 + W_2 + W_3 + W_4)$
= (44.1 + 8.3 + 8.2 + 137.9) = 198.5 kN

Resultant moment of the applied loads about the centre-line is given by: $(W_1e_1 + W_4e_4 - W_2e_2)$ **(Note:** e_3 is zero)

= [(44.1 × 63) + (137.9 × 49.2) − (8.3 × 76.3)]
≈ 8930 kNmm
(excluding 0.4*h* of wall 'W_3' vertical load = 190.3 kN)

Resultant eccentricity = $\dfrac{Resultant\ Moment}{Vertical\ Load}$ = $\dfrac{8930}{190.3}$

e_x ≈ 47 mm = $\dfrac{47}{255}t$ = 0.18 *t*

(where *t* is the effective thickness of the wall)

Clause 29.6(b)(4)

e_x ≤ 0.2*t* ∴ O.K.

Clause 27.3
Table 4

Partial Safety Factor for Material Strength (γ_m)
γ_m = 2.5   as before

Clause 28

Consideration of Slenderness of Walls and Columns
slenderness ratio (SR) = h_{ef}/t_{ef} ≤ 27

Clause 28.2.2

Horizontal Lateral Support
Since the concrete roof has a bearing length of at least one-half the thickness of the wall, enhanced resistance to lateral movement can be assumed.

Clause 28.3.1

Effective Height
h_{ef} = 0.75 × clear distance between lateral supports
= (0.75 × 3200) = 2400 mm

Clause 29.7

Effective Thickness
For grouted cavity walls the effective thickness can be taken as the actual thickness if the conditions specified are satisfied.

∴ t_{ef} = 255 mm

Contract : Warehouse Job Ref. No. : Example 2.7	Calcs. by : W.McK.
Part of Structure : Masonry Walls	Checked by :
Calc. Sheet No. : 8 of 10	Date :

References	Calculations	Output
	Slenderness ratio $= SR = \dfrac{2400}{255} = 9.41 \quad < 27$	
Table 7	Capacity Reduction Factor Linear interpolation between slenderness and eccentricity values is permitted when using Table 7.	

Table 7. Capacity reduction factor, β

Slender-ness ratio h_{ef}/t_{ef}	Eccentricity at top of wall, e_x			
	Up to 0.05t (see note 1)	0.1t	0.2t	0.3t
0	1.00	0.88	0.66	0.44
8	1.00	0.88	0.66	0.44
10	0.97	0.88	0.66	0.44
12	0.93	0.87	0.66	0.44

$\beta_{(SR=9.41;\ ex=0.18t)} \approx 0.7$

References	Calculations	Output
Clause 32.2.1	Design vertical load resistance $= \dfrac{\beta t f_k}{\gamma_m} \geq 198.5$ kN/m length	**Adopt bricks with unit strength ≥ 10 N/mm^2 and mortar type (i), (ii) (iii) or (iv).**
	$f_k = \dfrac{198.5 \times 2.5}{0.7 \times 255} = 2.8$ N/mm^2	
Table 2(a)	Use bricks with a compressive strength of 10 N/mm^2 and mortar designation either (i), (ii), (iii) or (iv).	**Note:** **28-day strength of the concrete infill must be consistent with the mortar type adopted.**
	(b) Solid wall supporting the front of the loading bay:	
	Design load due to the floor slab $= R_A$ $\qquad = [(259.2 + 52.4) - R_B]$ $\qquad = (311.6 - 137.9)$ $\qquad = 173.7$ kN/ metre length	
	Design load due to self-weight of $04h$ of wall $[1.4 \times (0.4 \times 1.0 \times 0.3275 \times 18)] = 3.3$ kN/m length	
	Total design load $= (173.7 + 3.3) = 176.6$ kN/metre length	

Contract : Warehouse Job Ref. No. : Example 2.7	Calcs. by : W.McK.
Part of Structure : Masonry Walls	Checked by :
Calc. Sheet No. : 9 of 10	Date :

References	Calculations	Output
Clause 31	Resultant moment of the applied loads about the centre-line is $= [(173.7) \times t/6] = 28.95t$ Resultant eccentricity $= \dfrac{Resultant\ Moment}{Vertical\ Load} = \dfrac{28.95t}{173.7}$ $e_x \approx 0.17t$ **Note:** this is very similar to the worst case of assuming ($t/6$).	
Clause 27.3	Partial Safety Factor for Material Strength (γ_m)	
Table 4	$\gamma_m = 2.5$ as before	
Clause 28	Consideration of Slenderness of Walls and Columns slenderness ratio (SR) $= h_{ef}/t_{ef} \leq 27$	
Clause 28.2.2	Horizontal Lateral Support Since the concrete roof has a bearing length of at least one-half the thickness of the wall, enhanced resistance to lateral movement can be assumed.	
Clause 28.3.1	Effective Height $h_{ef} = 0.75 \times$ clear distance between lateral supports $= (0.75 \times 1000) = 750$ mm	
Figure 3	Effective Thickness For solid walls the effective thickness can be taken as the actual thickness. $\therefore t_{ef} = 327.5$ mm	
Clause 28	Slenderness ratio $= SR = \dfrac{750}{327.5} = 2.3 < 27$	

References	Calculations	Output
	Contract : Warehouse **Job Ref. No. :** Example 2.7 **Part of Structure :** Masonry Walls **Calc. Sheet No. :** **10** of **10**	**Calcs. by :** W.McK. **Checked by :** **Date :**

References	Calculations	Output
Table 7	Capacity Reduction Factor Linear interpolation between slenderness and eccentricity values is permitted when using Table 7.	

Table 7. Capacity reduction factor, β

Slender-ness ratio h_{ef}/t_{ef}	Eccentricity at top of wall, e_x			
	Up to 0.05t (see note 1)	0.1t	0.2t	0.3t
0	1.00	0.88	0.66	0.44
6	1.00	0.66	0.88	0.44
8	1.00	0.88	0.66	0.44
10	0.97	0.88	0.66	0.44

$\beta_{(SR = 2.3;\; ex = 0.17t)} \approx 0.73$

References	Calculations	Output
Clause 32.2.1	Design vertical load resistance = $\dfrac{\beta\, t\, f_k}{\gamma_m} \geq 177$ kN/m length $f_k = \dfrac{177 \times 2.5}{0.73 \times 327.5} = 1.9$ N/mm^2	**Adopt bricks with unit strength ≥ 5 N/mm^2 and mortar type (i), (ii) (iii) or (iv).**

2.13   Example 2.8   Support For Fuel Tank

A small fuel tank is to be supported on two single-skin masonry walls each of which is constructed from solid concrete blocks. The load from the tank is distributed through timber beams as shown in Figure 2.47 and can be considered to be concentric on each wall.

Using the design data given determine a suitable blockwork/mortar combination which can be used for the support walls.

Design data:

Category of manufacturing control   normal
Category of construction control   normal
Solid concrete block dimensions: (440 mm long $\times$ 215 mm height $\times$ 190 mm thick)
Ultimate design load (i.e. including load factors)   200 kN

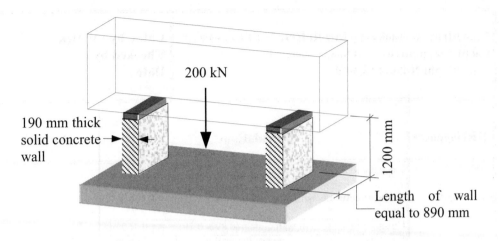

200 kN

190 mm thick solid concrete wall

1200 mm

Length of wall equal to 890 mm

Figure 2.47

2.13.1 Solution to Example 2.8

References	Calculations	Output
	Contract : **Tank Support** **Job Ref. No. :** **Example 2.8** **Calcs. by : W.McK.** **Part of Structure :** **Solid Concrete Block Walls** **Checked by :** **Calc. Sheet No. :** **1** of **2** **Date :**	

References	Calculations	Output
BS 5628 : Part 1	Structural use of unreinforced masonry	
	Length of wall = 890 mm	
	Design load = (200 / 0.890) = 224.7 kN/metre length	
Clause 27.3 Table 4 Clause 28	Partial Safety Factor for Material Strength (γ_m) γ_m = 3.5 Consideration of Slenderness of Walls and Columns slenderness ratio (SR) = h_{ef}/t_{ef} $\leq$ 27	
Clause 28.3.1	Effective Height In this case there is no resistance to lateral movement at the top of the wall. Despite the applied load being concentric it would be conservative to assume a more onerous effective length of 2.0h h_{ef} = 2.0 × clear height of the wall = (2.0 × 1200) = 2400 mm	
Figure 3	Effective Thickness For solid walls the effective thickness can be taken as the actual thickness. $\therefore t_{ef}$ = 190 mm	
Clause 28	Slenderness ratio = SR = $\dfrac{2400}{190}$ = 12.6 < 27	

References	Calculations	Output
	Contract : Tank Support **Job Ref. No. : Example 2.8** **Calcs. by : W.McK.** **Part of Structure :** Solid Concrete Block Walls **Checked by :** **Calc. Sheet No. : 2** of **2** **Date :**	

References	Calculations	Output
Table 7	Since the load is concentric use $e_x \le 0.05t$ $\therefore$ β = 0.92	
Clause 23.1.1	**Walls or columns with small plan area (i.e. area < 0.2 m^2)**	
	Cross-sectional area of wall = (0.89×0.19) = 0.169 m^2 Since the cross-sectional area of the wall is less than 0.2 m^2 the characteristic compressive strength f_k should be multiplied by:	
	$(0.7 + 1.5A)$ = $[0.7 + (1.5 \times 0.169)]$ = 0.95	
Clause 32.2.1	Design vertical load resistance = $\dfrac{\beta\, t\,(0.95 f_k)}{\gamma_{\mathrm m}} \ge$ 224.7 kN/m	
	$f_k \ge \dfrac{(224.7 \times 3.5)}{(0.92 \times 0.95 \times 190)}$ = 4.7 N/mm^2	
Table 2	aspect ratio of concrete block = $\dfrac{height}{least\ horizontal\ dimension}$ = $\dfrac{215}{190}$ = 1.13	
Clause 23.1.6	For solid concrete blocks with an aspect ratio between 0.6 and 2.0 f_k should be found by interpolation between the values given in Tables 2(b) and (d).	
Tables 2(b) & (d)	Assume blocks with a unit strength of 10 N/mm^2 and mortar designation is (ii) and interpolate between the two values given in Tables 2(b) and (d).	
	$f_k = \left[4.2 + (8.4 - 4.2) \times \dfrac{0.53}{1.4}\right] \approx$ 5.8 N/mm^2 > 4.7 N/mm^2 Adopt concrete blocks with unit compressive strength greater than or equal to 10 N/mm^2 and mortar type (ii).	**Adopt concrete blocks with unit strength $\ge$ 10 N/mm^2 with mortar type (ii)**

2.14   Concentrated Loads (Clause 34)

When considering relatively flexible members bearing on a wall the stress distribution due to the reaction is assumed to be triangular as shown in Figure 2.48.

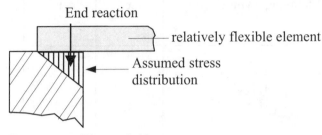

Figure 2.48

This is evident in Clause 31 when assessing the eccentricity, i.e. assuming '*that the load transmitted to a wall by a single floor or roof acts at one-third of the depth of the bearing area from the loaded face of the wall or loadbearing leaf.*'

In the case of loads (e.g. beam end reactions, column loads) which are applied to a wall through relatively stiff elements such as deep reinforced concrete beams, pad-stones or spreader beams as shown in Figure 2.49, provision is made in Clause 34 of the code for enhanced local bearing stresses.

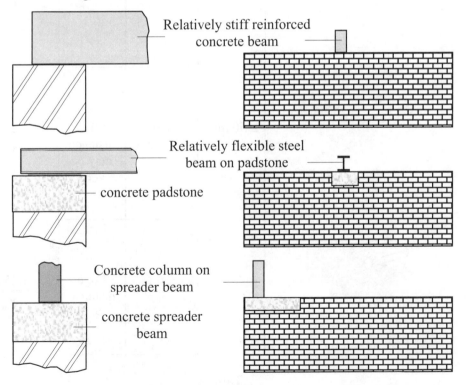

Figure 2.49

Design of Structural Masonry

Three types of bearing are considered in Figure 5 of the code and are illustrated in Table 2.7 and Table 2.8 in this text.

Bearing Type 1	x	y	z	**Local Design Strength**
	$\geq \frac{1}{2}\,t$	$\leq 3\,t$		$\dfrac{1.25 f_{k}}{\gamma_{m}}$
	$\geq \frac{1}{2}\,t$	$\leq 2\,t$		$\dfrac{1.25 f_{k}}{\gamma_{m}}$
	≥ 50 mm $\leq \frac{1}{2}\,t$	No restriction	Edge distance may be zero	$\dfrac{1.25 f_{k}}{\gamma_{m}}$
	$> \frac{1}{2}\,t$ $\leq t$	$\leq 6\,x$	Edge distance $\geq x$	$\dfrac{1.25 f_{k}}{\gamma_{m}}$

Table 2.7

Bearing Type 2	x	y	z	Local Design Strength
	≥ 50 mm $\leq \frac{1}{2}t$	$\leq 8x$	Edge distance $\geq x$	$\dfrac{1.5f_k}{\gamma_m}$
	$\geq \frac{1}{2}t$	$\leq 2t$		$\dfrac{1.5f_k}{\gamma_m}$
	$> \frac{1}{2}t$ $\leq t$	$\leq 4t$	Edge distance $\geq x$	$\dfrac{1.5f_k}{\gamma_m}$
Bearing Type 3				**Local Design Strength**
Spreader beam	The distribution of stress under the spreader beam should be derived from an acceptable elastic theory, e.g. (i) using simple elastic theory assuming a triangular stress block with zero tension, or (ii) using Timoshenko's analysis for elastic foundations (48)			$\dfrac{2.0f_k}{\gamma_m}$

Table 2.8

In '***Bearing Type 1***' an increase of 25% in the characteristic compressive strength (f_k), is permitted immediately beneath the bearing surface whilst in '***Bearing Type 2***' and '***Bearing Type 3***' 50% and 100% increases respectively are permitted.

    This local increase in strength at the bearing of a concentrated load is primarily due to the development of a triaxial state of stress in the masonry. Factors relating to the precise bearing details (e.g. relative dimensions of wall and bearing area, proximity of the bearing to the wall surface or wall end) also influence the ultimate failure load. The stress concentrations which occur in these circumstances rapidly disperse throughout the remainder of the masonry. It is normally assumed that this dispersal occurs at an angle of 45° as shown in Figure 2.50.

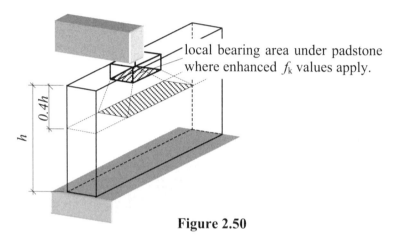

local bearing area under padstone where enhanced f_k values apply.

Figure 2.50

In Clause 34 the code requires that the strength of masonry be checked both locally under a concentrated load assuming the enhanced strength and at a distance of 0.4h below the level of the bearing where the strength should be checked in accordance with Clause 32. The effects of other loads, which are applied above the level of the bearing, should also be included in the calculated design stress. The assumed stress distributions for Bearing Types 1, 2 and 3 are illustrated in Figure 6 of the code and are shown in Figure 2.51.

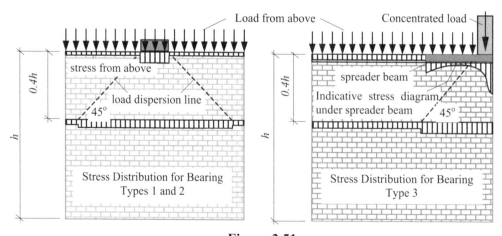

Figure 2.51

In Bearings Types 1 and 2 it is reasonable to assume a uniform stress distribution beneath the concentrated loads. Since the enhancement of f_k is partly due to the development of a triaxial state of stress beneath the load, in most cases it is assumed that restraint is provided at each side of the bearing. *It is important to recognise that strength enhancement cannot be justified in situations in which a minimum edge distance occurs on BOTH sides of a bearing.*

In Bearing Type 3 (i.e. a spreader beam), the stress distribution as indicated in Figure 2.51 clearly indicates a non-uniform distribution and the maximum value of stress should be considered. Two possible elastic solutions to estimate the maximum value, the Triangular Stress Block Method and Timoshenko's analysis for the bending of bars on elastic foundations are presented here.

2.14.1 Triangular Stress Block Method

This method is based on the '*middle-third*' rule as follows:

Consider a spreader beam of length 'L' distributing a concentrated load 'P' applied at an eccentricity 'e' from its mid-span axis as shown in Figure 2.52.

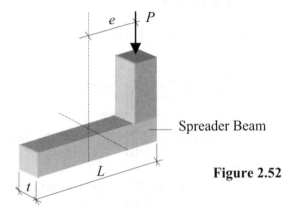

Spreader Beam

Figure 2.52

This applied load system can be represented by a concentric load 'P' in addition to a moment equal to 'Pe' as shown in Figure 2.53.

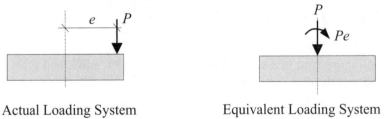

Actual Loading System   Equivalent Loading System

Figure 2.53

The equivalent load system is the sum of two load types one of which induces a uniform bearing pressure and one of which induces a moment pressure diagram as shown in Figure 2.54.

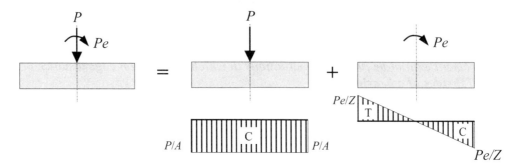

Uniform pressure diagram   Moment pressure diagram

where:

A is the area under the spreader beam $(t \times L)$

Z is the elastic section modulus about the axis of bending $(t \times L^2/6)$

Figure 2.54

The addition of these two pressure diagrams produces three alternative solutions:

(i)   where P/A > Pe/Z compression throughout

(ii)   where P/A = Pe/Z zero tension (limiting value)

(iii)   where P/A < Pe/Z tension and compression

Consider the case of zero tension   (i.e. a triangular stress block)

$$P/A = Pe/Z \quad \therefore e = Z/A = \frac{(t \times L^2)/6}{(t \times L)} = L/6$$

As indicated in Figure 2.55 this value defines a zone in which the load must lie to ensure that tension does not develop, i.e. 'the middle-third'.

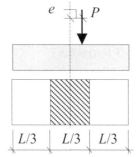

Figure 2.55

$L/3$ $L/3$ $L/3$

Assuming a triangular stress block, beneath the bearing with the line of action of the concentrated load passing through the centroid, enables the maximum compressive stress to be evaluated:

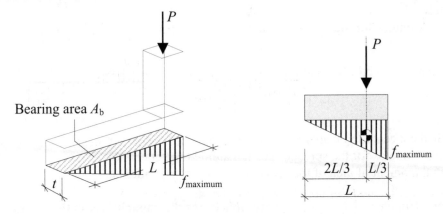

Figure 2.56

Considering vertical equilibrium: $\quad P = \frac{1}{2}(f_{maximum} \times A_b)$

$$\therefore f_{maximum} = (2P/A_b) = (2P/Lt)$$

2.14.2 *Timoshenko's Elastic Analysis*

An estimate of the maximum stress at the end of the spreader beam can be determined using the analysis presented by Timoshenko for the bending of bars on elastic foundations.

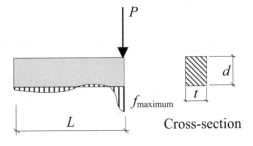

Cross-section

Figure 2.57

This can be used as follows:

$$f_{maximum} \approx (V/A_b)$$

where:

A_b is the bearing area and is equal to $(L \times t)$

V is defined as $(k \times \Delta)$ where:

k is a constant denoting the reaction when the deflection is unity and is given by:

$$k = \frac{A_b \, \delta \, E_b}{H}$$

where:
A_b is the bearing area under the spreader,
δ is the unit deflection,
H is the wall height,
E_b is Young's Modulus for the wall material (e.g. brickwork).

Δ is the calculated deflection beneath the load given by: $\quad \Delta = \dfrac{P}{2\beta^3 E_c I_z}$

where:
P is the applied concentrated load,
E_c is Young's Modulus for the spreader beam material (e.g. concrete),
I_z is the second moment of area of the cross-section of the spreader beam,

β is a constant defined by $\left(\dfrac{k}{4E_c I_z}\right)^{1/4}$ **(Note: this is not the capacity reduction factor)**

2.15   Example 2.9   Concentrated Load Due to Reinforced Concrete Beam

A 255 mm thick brickwork cavity wall in a masonry office block supports a reinforced concrete beam on its inner leaf as shown in Figure 2.58. Using the data provided determine a suitable brick/mortar combination.

Design data:

Category of manufacturing control	special
Category of construction control	normal
Ultimate load from above the level of the beam	21.0 kN/m
Ultimate load from the floor slab	8.0 kN/m
Ultimate end reaction from the beam	75 kN
Characteristic unit-weight of plaster	21.0 kN/m^3
Characteristic unit-weight of brickwork	18.0 kN/m^3
Assume that the walls are part of a braced structure	

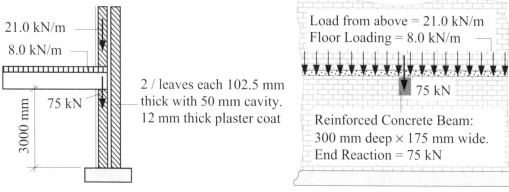

21.0 kN/m
8.0 kN/m
3000 mm
75 kN
2 / leaves each 102.5 mm thick with 50 mm cavity.
12 mm thick plaster coat

Load from above = 21.0 kN/m
Floor Loading = 8.0 kN/m
75 kN
Reinforced Concrete Beam:
300 mm deep × 175 mm wide.
End Reaction = 75 kN

Figure 2.58

2.15.1 Solution to Example 2.9

Contract : Office Block Job Ref. No. : Example 2.9 Part of Structure : Masonry Wall Under Beam Calc. Sheet No. : 1 of 5	Calcs. by : W.McK. Checked by : Date :

References	Calculations	Output
BS 5628 : Part 1	Structural use of unreinforced masonry	
	Self-weight of 12 mm thick plaster layer $\begin{aligned} &= (0.012 \times 21) \\ &= 0.25 \text{ kN/m}^2 \end{aligned}$	
	Self-weight of 102.5 mm thick cavity wall $\begin{aligned} &= (0.1025 \times 18) \\ &= 1.85 \text{ kN/m}^2 \end{aligned}$	
	Ultimate weight of wall and plaster $\begin{aligned} &= 1.4 \times (0.25 + 1.85) \\ &= 2.94 \text{ kN/m}^2 \end{aligned}$	
Clause 34	The bearing stress should be considered at two locations; section x-x for local stresses and section y-y for distributed stresses.	
Clause 27.3	Category of manufacturing control – special Category of construction control – normal	
Table 4	$\gamma_m = 3.1$	
	Consider section x-x	
Figure 5 (Table 2.8 of this text)	width of bearing x = 102.5 mm $\geq \tfrac{1}{2}t , \leq t$ length of bearing y = 175 mm $\leq 4t$ edge distance $z \geq x$	
	Assume Bearing Type 2 at location of the beam	
	Local strength $= \dfrac{1.5 f_k}{\gamma_m}$	

Contract : Office Block Job Ref. No. : Example 2.9 Part of Structure : Masonry Wall Under Beam Calc. Sheet No. : 2 of 5	Calcs. by : W.McK. Checked by : Date :

References	Calculations	Output
	  Design load from above = (29.0×0.175) = 5.1 kN Design load from beam = 75.0 kN Total design load at x-x = $(5.1 + 75.0)$ = 80.1 kN Bearing Type 2: Local resistance = $\left[\dfrac{1.5 f_k}{\gamma_m} \times A_b\right] \geq$ 80.1 kN A_b = (102.5×175) = 179.38×10^2 mm^2 $f_k \geq \left[\dfrac{80.1 \times 10^3 \times 3.1}{1.5 \times 179.38 \times 10^2}\right]$ = 9.2 N/mm^2	
Clause 23.1.2	The narrow brick wall factor applies $\therefore f_k$ required $= (9.2 / 1.15)$ $= 8.0$ N/mm^2 **Consider section y-y** $0.4h$ = (0.4×3000) = 1200 mm B = $[175 + (2 \times 0.4\,h)]$ = $[175 + (2 \times 1200)]$ = 2575 mm Load due to self-weight of $0.4h$ of wall = (1.2×2.9) = 3.5 kN/m length	

Contract : **Office Block** **Job Ref. No. : Example 2.9**	**Calcs. by : W.McK.**
Part of Structure : **Masonry Wall Under Beam**	**Checked by :**
Calc. Sheet No. : **3** of **5**	**Date :**

References	Calculations	Output
	Design load from above = 29.0 kN/m length Design load from beam = 75.0 kN This load is distributed over the length B = 2575 mm $\qquad\qquad = \dfrac{75.0}{2.575} = $ 29.1 kN/m length Design load at section y − y = (3.5 + 29.0 +29.1) $\qquad\qquad\qquad\qquad\quad\;$ = 61.6 kN/m length	
Clause 32.2.1	Design vertical load resistance/unit length $= \dfrac{\beta\, t\, f_k}{\gamma_m}$	
Table 4	γ_m = 3.1	
Clause 28	Consideration of Slenderness of Walls and Columns slenderness ratio (SR) $=\ h_{ef}/t_{ef}\ \le\ 27$	
Clause 28.2.2	Horizontal Lateral Support Since this structure has a concrete floor with a bearing length of at least one-half the thickness of the wall, enhanced resistance to lateral movement can be assumed.	
Clause 28.3.1	Effective Height h_{ef} = 0.75 × clear distance between lateral supports $\qquad$ = (0.75 × 3000) = 2250 mm	
Clause 28.4.1	Effective Thickness For cavity walls the effective thickness is as indicated in Figure 3 of the code and equal to the greatest of: (a) $2/3(t_1 + t_2)$ = $2/3(102.5 + 102.5)$ = 136.7 mm $\qquad$ or (b) $\qquad t_1$ = 102.5 mm $\qquad$ or (c) $\qquad t_2$ = 102.5 mm $\qquad \therefore\ t_{ef}$ = 136.7 mm Slenderness ratio = SR = $\dfrac{2250}{136.7}$ = 16.5 $\ <\ 27$	
Clause 31	Eccentricity Perpendicular to the Wall 29.0 kN 29.1 kN $\qquad e_x\ \approx\ \dfrac{(29.1\times t/6)}{61.6} = $ 0.08*t* 3.5 kN	

References	Calculations	Output

Contract : Office Block   Job Ref. No. : Example 2.9
Part of Structure :   Masonry Wall Under Beam
Calc. Sheet No. : 4 of 5

Calcs. by : W.McK.
Checked by :
Date :

Table 7

Capacity Reduction Factor

Table 7. Capacity reduction factor, β

Slender-ness ratio h_{ef}/t_{ef}	Eccentricity at top of wall, e_x			
	Up to 0.05t (see note 1)	0.1t	0.2t	0.3t
0	1.00	0.88	0.66	0.44
16	0.83	0.77	0.64	0.44
18	0.77	0.70	0.57	0.44

$\beta_{(SR = 16.5; \; ex = 0.08t)} \approx 0.8$

Clause 23.1.2

Narrow Brick Walls
When using standard format bricks to construct a wall one brick (i.e. 102.5 mm) wide, the values of f_k obtained from Table 2(a) can be multiplied by 1.15.

Clause 32.2.1

Design vertical load resistance = $\dfrac{\beta \, t \left(1.15 \times f_k\right)}{\gamma_m} \geq 61.6$ kN/m

$$f_k = \frac{(61.6 \times 3.1)}{(0.8 \times 102.5 \times 1.15)}$$

$$= 2.0 \text{ N/mm}^2$$

From the two values obtained i.e. 8.0 N/mm^2 and 2.1 N/mm^2 it is evident that the selection of unit strength is based on the local bearing strength at section x – x.

Since this is a localised problem the introduction of a bearing pad under the beam may result in a more economic solution.

Assume units with compressive strength of 10 N/mm^2 and Type (ii) mortar are to be used.

Table 2

$f_k = 4.2$ N/mm^2 ;

Local design resistance $= \left(\dfrac{1.5 f_k}{\gamma_m}\right) \times A_b \geq 61.6$ kN/m

$A_b \geq \left(\dfrac{61.6 \times 10^3 \times 3.1}{1.5 \times 4.2}\right) = 30.3 \times 10^3$ mm^2

Contract : Office Block Job Ref. No. : Example 2.9 Part of Structure : Masonry Wall Under Beam Calc. Sheet No. : 5 of 5	Calcs. by : W.McK. Checked by : Date :

References	Calculations	Output
	Min. length of bearing required $= \dfrac{30.3\times10^3}{102.5} = $ 296 mm For Type 2 bearing with $\tfrac{1}{2}t \leq x \leq t$ the length 'y' $\leq 4t$ $\therefore y \leq (4 \times 102.5) = $ 410 mm > required length of 296 mm Adopt a bearing pad: (assume 45° dispersion of load) (320 mm long × 102.5 mm wide × 150 mm thick)	**Adopt:** **Units with compressive strength ≥ 10.0 N/mm^2 in mortar Type (ii).** **Use concrete padstone 320 mm long × 102.5 mm wide × 150 mm thick**

2.16   Example 2.10   Concentrated Load Due to Reinforced Concrete Column

A masonry wall constructed from solid concrete blocks is required to support a floor slab and a series of columns as shown in Figure 2.59. Using the data provided check the suitability of the wall to support a 10% increase in the floor loading.

Design data:

Assume the category of manufacturing control	normal
Assume the category of construction control	normal
Ultimate design load in the column	120.0 kN
Ultimate load from the floor slab (before the increase in loading)	30.0 kN/m length
Solid concrete blocks:	450 mm long × 215 mm high × 190 mm wide
Characteristic strength of unit	15.0 N/mm^2
Mortar designation	Type (iii)
Characteristic unit weight of wall	24.0 kN/m^3
Assume that the walls are part of a braced structure	

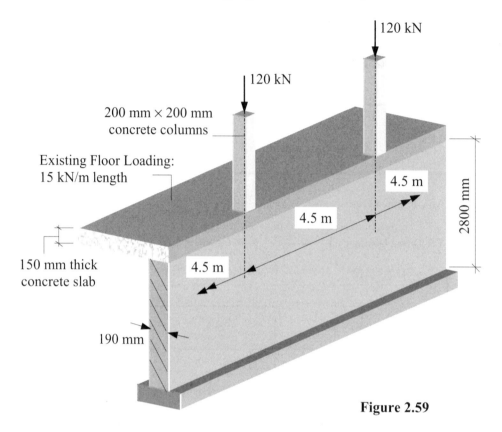

Figure 2.59

2.16.1   Solution to Example 2.10

References	Calculations	Output
Contract : Column Support **Job Ref. No. :** Example 2.10 **Part of Structure :** Masonry Wall **Calc. Sheet No. : 1** of **3**		**Calcs. by : W.McK.** **Checked by :** **Date :**
BS 5628 : Part 1	Structural use of unreinforced masonry Design load in the columns = 120 kN Design load on floor = (1.1×15) = 16.5 kN/m	
Clause 34	The bearing stress should be considered at two locations: section x-x for local stresses and section y-y for distributed stresses. (see page 2 of the calculation sheets). The column load disperses throughout the depth of the floor slab resulting in a bearing length of 500 mm at the top of the wall. This can be considered as Bearing Type 2. i.e. length $\leq$ $4t$ $[(4 \times 190)$ = 760 mm] width $>$ $\frac{1}{2}t$, $\leq$ t	

Contract : Column Support Job Ref. No. : Example 2.10	Calcs. by : W.McK.
Part of Structure : Masonry Wall	Checked by :
Calc. Sheet No. : **2** of **3**	Date :

References	Calculations	Output
	Consider section x-x	
	Applied load $= \left(16.5 + \dfrac{120}{0.5}\right) = 256.5$ kN/m length	
Figure 5	Local resistance $= \left[\dfrac{1.5 f_k}{\gamma_m} \times A_b\right] \geq 256.5$ kN/m length	**Local bearing strength is adequate**
Clause 27.3	Category of manufacturing control – normal	
	Category of construction control – normal	
Table 4	$\gamma_m = 3.5$	
Clause 23.1.6	Aspect ratio of concrete blocks $= \dfrac{215}{190} = 1.13$	
	Since $0.6 < \dfrac{height}{least\ horizontal\ dimension} < 2.0$ use	
	interpolation between Tables 2(b) and (d) to determine f_k	
Table 2(b)	Unit strength $= 15.0$ N/mm^2 and Mortar Type (iii)	
	Characteristic compressive strength $f_k = 4.1$ N/mm^2	
Table 2(d)	$f_k = 8.2$ N/mm^2	

References	Calculations	Output
	Contract : Column Support **Job Ref. No. :** Example 2.10 **Part of Structure :** Masonry Wall **Calc. Sheet No. : 3** of **3**	**Calcs. by :** W.McK. **Checked by :** **Date :**

References	Calculations	Output
	$f_k = \left[5.0 + \left(5.0 \times \dfrac{0.53}{1.4} \right) \right]$ $= 6.8 \text{ N/mm}^2$	
	Local resistance $= \left[\dfrac{1.5 f_k}{\gamma_m} \times A_b \right]$	
	$A_b = (190 \times 500) = (95 \times 10^3) \text{ mm}^2$	
	Local resistance $= \left[\dfrac{1.5 \times 6.8}{3.5} \left(95 \times 10^3 \right) \right] \Big/ 10^3$	
	$\qquad = 276 \text{ kN/m length} \quad > 256.6 \text{ kN/m length}$	
	Consider section y -y	
	B $= 2740 \text{ mm}; \qquad 0.4h = (0.4 \times 2.800) = 1.12 \text{ m}$	
	Ultimate load due to self-weight of $0.4h$ of wall:	
	$\qquad = 1.4 \, (1.12 \times 0.19 \times 24.0) \quad = \quad 7.2 \text{ kN/m length}$	
	Design load due to column $= 120 \text{ kN} = \dfrac{120}{2.74} = 43.8 \text{ kN/m}$	
	Design load due to floor $= 16.5 \text{ kN/m}$	
	Total design load $= (7.2 + 43.8 + 16.5) = 67.5 \text{ kN/m}$	
Clause 32.2.1	Design vertical load resistance/unit length $= \dfrac{\beta\, t\, f_k}{\gamma_m}$	
Clause 28.3.1	Effective Height $h_{ef} = 0.75 \times$ clear distance between lateral supports $\qquad = (0.75 \times 2800) = 2100 \text{ mm}$	
Clause 28.4.1	Effective Thickness $= t_{ef} = 190 \text{ mm}$	
	Slenderness ratio $= \text{SR} = \dfrac{2100}{190} = 11.1 \quad < 27$	
Table 7	Eccentricity $\leq 0.05t$ Capacity reduction factor $= 0.94$	
Clause 32.2.1	Design vertical load resistance/unit length $= \dfrac{0.94 \times 190 \times 6.8}{3.5}$	**Existing wall is adequate to resist the 10% increase in floor loading.**
	$\qquad = 347 \text{ kN/m}$	
	$\qquad \gg 67.5 \text{ N/mm}^2$	

2.17   Example 2.11   Roof Truss Support Wall

The roof of a sports hall comprises a series of roof trusses supported on the inner leaf of a stiffened cavity wall as shown in Figure 2.60. Using the design data given determine a suitable brick mortar combination.

Design data:

Assume the category of manufacturing control	normal
Assume the category of construction control	normal
Characteristic dead load due to trusses and roof loading	2.5 kN/m^2
Characteristic imposed load due to roof loading	1.5 kN/m^2
Characteristic unit weight of brickwork	18.0 kN/m^3
Truss spacing	7.6 m

Assume that the walls are part of a braced structure.
Lateral wind loading is not being considered in this example.

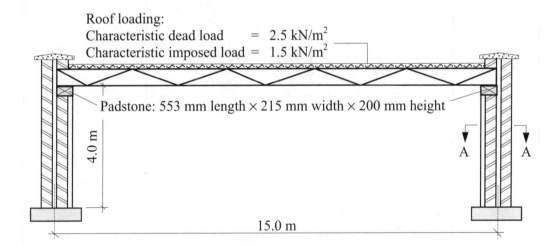

Roof loading:
Characteristic dead load   =   2.5 kN/m^2
Characteristic imposed load =   1.5 kN/m^2

Padstone: 553 mm length × 215 mm width × 200 mm height

4.0 m

15.0 m

A   A

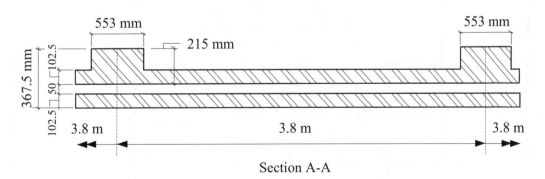

553 mm   215 mm   553 mm

367.5 mm   102.5   50   102.5

3.8 m   3.8 m   3.8 m

Section A-A

Figure 2.60

2.17.1 *Solution to Example 2.11*

Contract : Sports Hall Job Ref. No. : Example 2.11	Calcs. by : W.McK.
Part of Structure : Roof Truss Support Walls	Checked by :
Calc. Sheet No. : 1 of 3	Date :

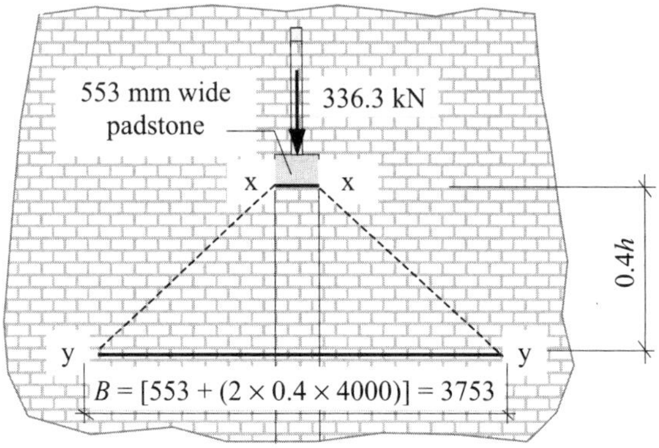

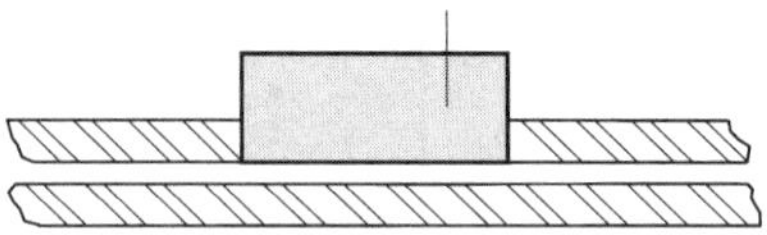

References	Calculations	Output
BS 5628 : Part 1	Structural use of unreinforced masonry	

Design dead load due to truss and roof loading:
$$= 1.4(2.5 \times 15.0 \times 7.6) = 399.0 \text{ kN}$$
Design imposed load due to truss and roof loading:
$$= 1.6(1.5 \times 15.0 \times 7.6) = 273.6 \text{ kN}$$
Total design load on truss $= (399.0 + 273.6) = 672.6 \text{ kN}$
End reaction on brickwork pier of inner leaf $= (0.5 \times 676.6)$
$$= 336.3 \text{ kN}$$

553 mm wide padstone — 336.3 kN

0.4*h*

$B = [553 + (2 \times 0.4 \times 4000)] = 3753$

Consider section x – x
Padstone details:

553 mm long $\times$ 215 mm wide

$x = 215$ mm $(= t)$, $y = 553$ mm $(< 4t)$
z (edge distance) $> x$

Figure 5 — This conforms with Bearing Type 2

Applied bearing stress under the bearing pad $= \dfrac{336.3 \times 10^3}{(553 \times 215)}$
$$= 2.83 \text{ N/mm}^2$$

Local design strength $= \dfrac{1.5 f_k}{\gamma_m} \geq 2.83 \text{ N/mm}^2$

Clause 27.3 — Category of construction/manufacturing control: normal/normal
Table 4 — $\gamma_m = 3.5$

Contract : Sports Hall	Job Ref. No. : Example 2.11	Calcs. by : W.McK.
Part of Structure : Roof Truss Support Walls		Checked by :
Calc. Sheet No. : 2 of 3		Date :

References	Calculations	Output
	$f_k \geq \dfrac{2.83 \times 3.5}{1.5} = 6.6 \text{ N/mm}^2$ **Consider section y – y** $B = [553 + (2 \times 1600)] = 3753 \text{ mm}$ Area of inner leaf at section y – y = $[(553 \times 112.5) + (3753 \times 102.5)] = 446.9 \times 10^3 \text{ mm}^2$ Design load due to self-weight of $0.4h$ of wall = $1.4[18.0 \times (0.4 \times 4.0) \times 446.9 \times 10^3]/10^6 = 18.0 \text{ kN}$ Total design load $= (336.3 + 18.0) = 354.3 \text{ kN}$ $= \dfrac{354.3}{3.753} = 94.4 \text{ kN/m}$ Equivalent wall thickness: $t_\text{equivalent} = \dfrac{A_1 + A_2}{L_p}$ $A_1 = (553 \times 112.5) = 62.2 \times 10^3 \text{ mm}^2$ $A_2 = (3800 \times 102.5) = 389.5 \text{ mm}^2$ $L_p = 3800 \text{ mm}$ $t_\text{equivalent} = \dfrac{(62.2 + 389.5) \times 10^3}{3800} = 118.9 \text{ mm}$	

see Section 2.5

Contract : Sports Hall Job Ref. No. : Example 2.11	Calcs. by : W.McK.
Part of Structure : Roof Truss Support Walls	Checked by :
Calc. Sheet No. : **3** of **3**	Date :

References	Calculations	Output
Clause 28.4.2 Table 5	Effective thickness $t_{ef} = t \times K$ $\dfrac{pier\ spacing}{pier\ width} = \dfrac{3800}{553} = 6.9$ $\dfrac{pier\ thickness}{actual\ thickness} = \dfrac{215}{102.5} = 2.1$ *Stiffness Coefficient for Wall Stiffened by Piers (K)* **Table 5. Stiffness coefficient for walls stiffened by piers** <table><tr><td>Ratio of pier spacing (centre to centre) to pier width</td><td colspan="3">Ratio t_p/t of pier thickness to actual thickness of wall to which it is bonded</td></tr><tr><td></td><td>1</td><td>2</td><td>3</td></tr><tr><td>6</td><td>1.0</td><td>1.4</td><td>2.0</td></tr><tr><td>10</td><td>1.0</td><td>1.2</td><td>1.4</td></tr><tr><td>20</td><td>1.0</td><td>1.0</td><td>1.0</td></tr></table>NOTE. Linear interpolation between the values given in table 5 is permissible, but not extrapolation outside the limits given. Interpolation from Table 5 gives $K = $ 1.41 Effective thickness = (1.41×102.5) = 144.5 mm	
Clause 28.3.1 Clause 28.0 Table 7	Effective height = (0.75×4000) = 3000 mm Slenderness ratio SR = $\dfrac{3000}{144.5}$ = 20.8 < 27 Eccentricity $\leq$ $0.05t$ $\beta = \left[0.7 - \left(0.08 \times \dfrac{0.8}{2.0}\right)\right]$ = 0.67 Design strength = $\dfrac{\beta\, t\, f_k}{\gamma_m}$ $\geq$ 94.4 kN/m $f_k \geq \dfrac{94.4 \times 3.5}{0.67 \times 118.9}$ = 4.15 N/mm²	Select a combination from Table 5(a) such that $f_k \geq 6.6$ N/mm² e.g. unit strength: 27.5 N/mm² mortar type (iii)

2.18 Example 2.12 Concentrated Load on Spreader Beam

The load from a reinforced concrete column in a warehouse building is transmitted to a single-leaf brickwork wall through a spreader beam as shown in Figure 2.61. Using the design data given check the suitability of the spreader beam indicated.

Design data:

Assume the category of manufacturing control	normal
Assume the category of construction control	special
Ultimate design load in the column	200.0 kN
Ultimate design load from above and floor slab	75 kN/m
Characteristic strength of unit (standard format bricks)	20.0 N/mm^2
Mortar designation	Type (ii)
Wall thickness	215 mm
Characteristic unit weight of brickwork	18.0 kN/m^3
Modulus of elasticity of brickwork (E_b)	900f_k
Modulus of elasticity of concrete spreader beam (E_c)	25×10^6 kN/m^2

Assume that the walls are part of a braced structure.

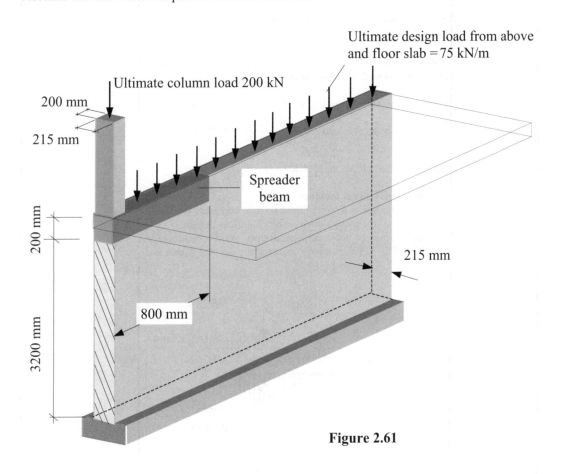

Ultimate design load from above and floor slab = 75 kN/m

Ultimate column load 200 kN

200 mm

215 mm

Spreader beam

200 mm

215 mm

800 mm

3200 mm

Figure 2.61

2.18.1 Solution to Example 2.12

References	Calculations	Output
	Contract : Warehouse　**Job Ref. No. :** Example 2.12　**Calcs. by :** W.McK. **Part of Structure :**　Masonry Wall Under Spreader　**Checked by :** **Calc. Sheet No. : 1** of **5**　**Date :**	

References	Calculations	Output
BS 5628 : Part 1	Structural use of unreinforced masonry The spreader beam under the column is Bearing Type 3. Two methods of estimating the maximum stress under the beam are described in sections 2.14.1 and 2.14.2. they are; (i) the triangular stress block method　　　　and (ii) Timoshenko's elastic analysis for bars on elastic foundations The use of both methods is illustrated in the solution to this example. The maximum stress due to the column load must be added to any stress induced from any other level. Ultimate load from above and floor slab　　=　75 kN/m length Stress induced by this load = $\dfrac{75\times10^3}{1000\times215}$ = 0.35 N/mm^2 Ultimate column load =　200 kN **(i) Triangular stress method:** Assuming the line of action of the column load passes through the centroid of the triangular stress block implies that the spreader beam need not be any longer than 300 mm. In this case since the beam is 800 mm long, theoretically, tension occurs as indicated. 200 mm　　P = 200 kN t = 215 mm Theoretical tension zone f_{maximum} 100 mm L $(3 \times 100) = 300$ mm	

Contract : Warehouse Job Ref. No. : Example 2.12	Calcs. by : W.McK.
Part of Structure : Masonry Wall Under Spreader	Checked by :
Calc. Sheet No. : 2 of 5	Date :

References	Calculations	Output
See section 2.14.1	The maximum stress can be estimated using vertical equilibrium.	

See section 2.14.1

The maximum stress can be estimated using vertical equilibrium.

$$P = \text{(average stress} \times \text{area)} = \left(\frac{f_{\text{maximum}}}{2} \times L \times t \right)$$

$$f_{\text{maximum}} = \frac{2P}{Lt} = \frac{(2 \times 200 \times 10^3)}{(300 \times 215)} = 6.2 \ \text{N/mm}^2$$

This value must be added to that obtained from the floor loading.

Total stress $f_{\text{total}} = (0.35 + 6.2) = 6.55 \ \text{N/mm}^2$

See section 2.14.2 **(ii) Timoshenko's elastic analysis:**

200 kN

200 mm

215 mm

f_{maximum}

800 mm

Cross-section

The maximum stress can be estimated using

$$f_{\text{maximum}} = \frac{V}{A_b} \quad \text{where:}$$

V is defined as $(k \times \Delta)$ and

$$k = \frac{A_b \ \delta \ E_b}{H}; \quad \Delta = \frac{P}{2\beta^3 E_c I_z}; \quad \beta = \left(\frac{k}{4 E_c I_z} \right)^{1/4}$$

where:
A_b is the bearing area,
δ is unit deflection,
k is constant,
H is the wall height,
E_b is Young's Modulus for the wall material,
E_c is Young's Modulus for the spreader beam material,
I_z is the second moment of area of the spreader beam,
β is a constant **(Note: this is not the capacity reduction factor)**
P is the applied load.

References	Calculations	Output
	$A_b = (L \times t) = (0.8 \times 0.215) = 0.172 \text{ m}^2$, $\delta = 1.0$, $H = 3.2 \text{ m}$, E_b is the Young's modulus of elasticity for the brickwork. This parameter is very variable and a value of $900 f_k$ is assumed here.	
BS 5628 : Part 2 Clause 7.4.1.7		
Table 2(a)	Unit strength $= 20 \text{ N/mm}^2$ and mortar is type (ii) $\therefore f_k = 6.4 \text{ N/mm}^2$ $E_b = (900 \times 6.4) = 5760 \text{ N/mm}^2 = (5.76 \times 10^6 \text{ kN/m}^2)$ $E_c = 25 \times 10^6 \text{ kN/m}^2$ $I_z = \left[\dfrac{0.215 \times 0.2^3}{12} \right] = 143.3 \times 10^{-6} \text{ m}^4$ $k = \left[\dfrac{0.172 \times 1.0 \times 5.76 \times 10^6}{3.2} \right] = 309.6 \times 10^3 \text{ kN/m}$ $\beta = \left[\dfrac{309.6 \times 10^3 \times 10^6}{4 \times 25 \times 10^6 \times 143.3} \right]^{\frac{1}{4}} = 2.16$ $\Delta = \left[\dfrac{200 \times 10^6}{2 \times 2.16^3 \times 25 \times 10^6 \times 143.3} \right] = 2.77 \times 10^{-3} \text{ m}$ $P = 200 \text{ kN}$ $V = (k \times \Delta) = (309.6 \times 10^3 \times 2.77 \times 10^{-3}) = 858 \text{ kN}$ $f_{maximum} = \left[\dfrac{858 \times 10^3}{800 \times 215} \right] = 4.98 \text{ N/mm}^2$ This value must be added to that obtained from the floor loading. Total stress $\quad f_{total} = (0.35 + 4.98) = 5.33 \text{ N/mm}^2$	
Figure 5 Table 4	Local strength $= \dfrac{2 f_k}{\gamma_m} \geq f_{total}$ Category of manufacturing control: normal Category of construction control: special $\gamma_m = 2.8$ $\dfrac{2 f_k}{\gamma_m} = \dfrac{2 \times 6.4}{2.8} = 4.57 \text{ N/mm}^2 \; < \; f_{total}$	

Contract : Warehouse   **Job Ref. No. :** Example 2.12   **Calcs. by :** W.McK.
Part of Structure : Masonry Wall Under Spreader   **Checked by :**
Calc. Sheet No. : 3 of **5**   **Date :**

Contract : Warehouse Job Ref. No. : Example 2.12	Calcs. by : W.McK.
Part of Structure : Masonry Wall Under Spreader	Checked by :
Calc. Sheet No. : 4 of 5	Date :

References	Calculations	Output
	Since the local strength is less than the total stress calculated a larger spreader beam is required. e.g. try a 1200 mm long × 215 mm wide beam. $A_b = (1.2 \times 0.215) = 0.258 \text{ m}^2$ Using proportion determine the revised values of the variables. $k = \left[\left(309.6 \times 10^3\right) \times \dfrac{0.258}{0.172} \right] = 464.4 \times 10^3$ $\beta = \left[2.16 \times \left(\dfrac{464.4}{309.6} \right)^{\frac{1}{4}} \right] = 2.39$ $\Delta = \left[\left(2.77 \times 10^{-3}\right) \times \left(\dfrac{2.16}{2.39} \right)^3 \right] = 2.04 \times 10^{-3} \text{ m}$ $V = (k \times \Delta) = (464.4 \times 10^3 \times 2.04 \times 10^{-3}) = 947.38 \text{ kN}$ $f_{\text{maximum}} = \left[\dfrac{947.38 \times 10^3}{1200 \times 215} \right] = 3.67 \text{ N/mm}^2$ Total stress $f_{\text{total}} = (0.35 + 3.67) = 4.02 \text{ N/mm}^2$ $< 4.57 \text{ N/mm}^2$ **Consider section y – y** $0.4h = [0.4 \times (3.2 + 0.2)] = 1.36 \text{ m}$ $B = (0.2 + 1.36) = 1.56 \text{ m}$ Self-weight of $04h$ of wall $= (0.4 \times 1.36 \times 1.56 \times 0.215 \times 18)$ $= 3.3 \text{ kN}$	Adopt a spreader beam: **1200 mm long ×** **215 mm wide ×** **200 mm high**

References	Calculations	Output
	$$f_{total} = \left[0.35 + \left(\frac{3.3 \times 10^3}{215 \times 1560}\right) + \left(\frac{200 \times 10^3}{1560 \times 215}\right)\right] = 0.96 \text{ N/mm}^2$$	
Clause 28.3 Clause 28.4	$h_{ef} = (0.75 \times 3200) = 2400$ mm $t_{ef} = 215$ mm	
Clause 28.1	Slenderness ratio $SR = \dfrac{2400}{215} = 11.2 < 27$	
	Eccentricity $e_x \leq 0.05t$	
Table 7	Capacity reduction factor $= \beta = 0.94$	**215 mm thick wall is acceptable**
	$\dfrac{\beta f_k}{\gamma_m} = \dfrac{0.94 \times 6.4}{2.8} = 2.14 \text{ N/mm}^2 > f_{total}$ (0.96 N/mm²)	

Contract : Warehouse   **Job Ref. No. :** Example 2.12   **Calcs. by :** W.McK.
Part of Structure :   Masonry Wall Under Spreader   **Checked by :**
Calc. Sheet No. : 5 of **5**   **Date :**

2.19   Columns   (Clause 32.2.2)

An isolated, axially loaded masonry element in which the width is not greater than four times the thickness is defined as a column as indicated in Figure 3 of the code. In addition the masonry between any two openings in a wall is, by definition, a column as illustrated in section 2.1.5, Figure 2.10. In most cases columns are solid but they may have a cavity as shown in Figure 2.62.

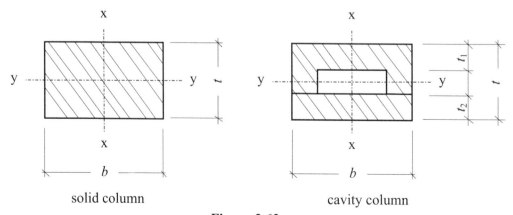

solid column   cavity column

Figure 2.62

In cavity columns [$b \leq$ (4 × the overall thickness)] when both leaves are loaded and [$b \leq$ (4 × thickness of the loaded leaf)] when only one leaf is loaded when defining a column.

The design of axially loaded columns is very similar to that for walls with additional consideration being given to the possibility of buckling about both the x –x axis and the y-y axis. (In walls support is provided in the longitudinal direction by adjacent material and only the y-y axis need be considered.) In Clause 32.2.2 the design vertical load resistance of a rectangular column is given by:

$$\frac{\beta \, b t \, f_k}{\gamma_m}$$

where:
b is the width of the column,
t is the thickness of the column,
β, f_k and γ_m are as before.

The value of the capacity reduction factor β is determined by considering four possible cases of eccentricity and slenderness as described in section 2.1.8. The effective height of columns is defined in Clause 28.3.1.2 and Clause 28.3.1.3 (see section 2.1.5).

Piers in which the thickness 't_p', is greater than 1.5 × the thickness of the wall of which they form a part, should be treated as columns for effective height considerations.

The effective height of columns should be taken as the distance between lateral supports or twice the height of the column in respect of a direction in which lateral support is not provided as indicated in Clause 28.3.1.2. (**Note:** *enhanced* resistance to lateral movement applies to walls, not columns.)

As indicated in Clause 28.2.2.2 resistance to lateral movement may be assumed where:

(a) *' floors or roofs of any form of construction span on to the wall or column, from both sides at the same level;*

(b) *an in-situ concrete floor or roof, or a precast concrete floor or roof giving equivalent restraint, irrespective of the direction of the span, has a bearing of at least one-half the thickness of the wall or inner leaf of a cavity wall or column on to which it spans but in no case less than 90 mm;*

(c) *in the case of houses of not more than three storeys, a timber floor spans on to a wall from one side and has a bearing of not less than 90 mm.*

Preferably, columns should be provided with lateral support in both horizontal directions.'

Typical restraint conditions and their respective slenderness ratio are illustrated in Figure 2.63.

Design of Structural Masonry

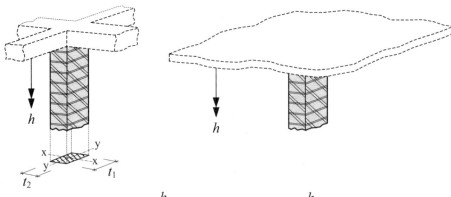

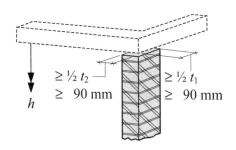 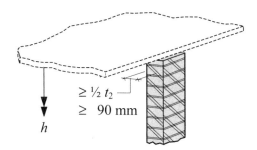

Slenderness ratios $\qquad$ $SR_{xx} = \dfrac{h_{ef\,xx}}{t_{ef\,xx}}$ $\quad$ and $\quad$ $SR_{yy} = \dfrac{h_{ef\,yy}}{t_{ef\,yy}}$

Slenderness ratios $\qquad$ $SR_{xx} = \dfrac{h_{ef\,xx}}{t_{ef\,xx}}$ $\quad$ and $\quad$ $SR_{yy} = \dfrac{h_{ef\,yy}}{t_{ef\,yy}}$

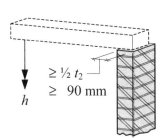 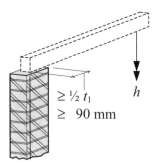

Slenderness ratios $\quad$ $SR_{xx} = \dfrac{2h_{ef\,xx}}{t_{ef\,xx}}$ $\qquad\qquad$ $SR_{xx} = \dfrac{h_{ef\,xx}}{t_{ef\,xx}}$

$\qquad\qquad$ and $\quad$ $SR_{yy} = \dfrac{h_{ef\,yy}}{t_{ef\,yy}}$ $\qquad\qquad$ $SR_{yy} = \dfrac{2h_{ef\,yy}}{t_{ef\,yy}}$

Figure 2.63

2.20 Example 2.13 Eccentrically Loaded Column

The cross-section of an eccentrically loaded column is shown in Figure 2.64. Using the design data given, determine the maximum value of the load P which can be applied in each of the cases (i) to (iv) indicated.

Design data:

Assume the category of manufacturing control	special
Assume the category of construction control	normal
Characteristic strength of unit (standard format bricks)	35.0 N/mm^2
Mortar designation	Type (ii)
Effective height about the y-y axis $(h_{ef\,yy})$	2500 mm
Effective height about the x-x axis $(h_{ef\,xx})$	5000 mm

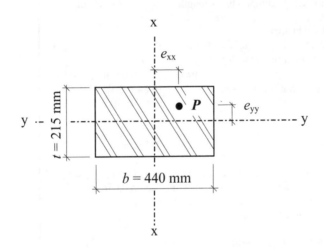

Case	e_{xx} (mm)	e_{yy} (mm)
(i)	15.0	8.0
(ii)	20.0	25.0
(iii)	30.0	6.0
(iv)	40.0	20.0

Figure 2.64

2.20.1 Solution to Example 2.13

Contract : Column Job Ref. No. : Example 2.13 Part of Structure : Eccentrically Loaded Column Calc. Sheet No. : 1 of 5	Calcs. by : W.McK. Checked by : Date :

References	Calculations	Output
BS 5628 : Part 1	Structural use of unreinforced masonry	
Clause 32.2.2	Design vertical load resistance $= \dfrac{\beta\, b\, t\, f_k}{\gamma_m}$	
Table 2(a)	Characteristic compressive strength $f_k = 9.4 \text{ N/mm}^2$	
Clause 23.1.1	Small plan area: Column area $= (0.215 \times 0.44) = 0.095 \text{ m}^2 \quad < 0.2 \text{ m}^2$ Use the multiplying factor $= (0.7 + 1.5A)$ $\qquad\qquad = [0.7 + (1.5 \times 0.095)] = 0.84$ Modified compressive strength $= (0.84 \times 9.4)$ $\qquad\qquad\qquad = 7.9 \text{ N/mm}^2$	
Table 4	$\gamma_m = 3.1$ $b = 440 \text{ mm} \qquad\qquad t = 215 \text{ mm}$ The capacity reduction factor β is determined according to the actual eccentricities e_{xx} and e_{yy} with respect to $0.05b$ and $0.05t$ as indicated in Clause 32.2.2 (a), (b), (c) and (d). (See section 2.1.8 of this text). $0.05b = (0.05 \times 440) = 22.0 \text{ mm}$ $0.05t = (0.05 \times 215) = 10.8 \text{ mm}$ **Case (i):** $e_{xx} = 15.0 \text{ mm} \quad < \quad 0.05b$ $e_{yy} = 8.0 \text{ mm} \quad < \quad 0.05t$	
Clause 32.2.2 (a)	Use Table 7 Slenderness ratio $= \dfrac{h_{ef}\ (relative\ to\ the\ minor\ axis)}{t_{ef}\ (based\ on\ column\ thickness\ t)}$ $SR = \dfrac{2500}{215} = 11.6; \qquad e_x \leq 0.05t$	
Table 7	Using interpolation $\beta = 0.94$ $P \leq \dfrac{\beta\, b\, t\, f_k}{\gamma_m} = \dfrac{0.94 \times 440 \times 215 \times 7.9}{3.1 \times 10^3} = 226.6 \text{ kN}$	$P \leq$ **226.6 kN**

Contract : Column Job Ref. No. : Example 2.13 Part of Structure : Eccentrically Loaded Column Calc. Sheet No. : 2 of 5	Calcs. by : W.McK. Checked by : Date :

References	Calculations	Output
	Case (ii): $e_{xx} = 20.0$ mm $< 0.05b$ $e_{yy} = 25$ mm $> 0.05t$	
Clause 32.2.2 (a)	Use Table 7 Slenderness ratio $= \dfrac{h_{ef} \text{ (relative to the minor axis)}}{t_{ef} \text{ (based on column thickness } t)}$ $SR = \dfrac{2500}{215} = 11.6;$ Use eccentricity appropriate to the minor axis: $e_x = e_{yy} = \dfrac{25.0}{215} t = 0.12t$	
Table 7	Using interpolation $\beta = 0.85$ $P \leq \dfrac{\beta \, b \, t \, f_k}{\gamma_m} = \dfrac{0.85 \times 440 \times 215 \times 7.9}{3.1 \times 10^3} = 204.9$ kN	$P \leq 204.9$ kN
	Case (iii): $e_{xx} = 30.0$ mm $> 0.05b$ $e_{yy} = 6.0$ mm $< 0.05t$ There are two options given in the code to determine the value for β : (a) Use the eccentricity related to e_{xx} , the slenderness appropriate to the minor axis and Table 7 or (b) Use Appendix B. ***Consider option (a)*** $e_{xx} = \dfrac{30.0}{440} b = 0.07b$ Slenderness ratio $= \dfrac{h_{ef} \text{ (relative to the minor axis)}}{t_{ef} \text{ (based on column thickness } t)}$ $SR = \dfrac{2500}{215} = 11.6$	

References	Calculations	Output

Contract : Column   **Job Ref. No. :** Example 2.13
Part of Structure :   Eccentrically Loaded Column
Calc. Sheet No. : 3 of 5

Calcs. by : W.McK.
Checked by :
Date :

References	Calculations	Output

Table 7

Using interpolation β = 0.91

$$P \le \frac{\beta\,b\,t\,f_k}{\gamma_m} = \frac{0.91\times440\times215\times7.9}{3.1\times10^3} = 219.4 \text{ kN}$$

Output: $P \le$ **219.4 kN**

Appendix B
Equation 4

Consider option (b):
β = $1.1[1 - (2e_m /t)]$

(see section 2.1.7) e_m is the larger of e_x and e_t but should not be taken as less than $0.05t$,
where:
e_x is the eccentricity at the top of the column,
e_t = $(0.6e_x + e_a)$ and
$$e_a = t\left[\frac{1}{2400}\left(h_{ef}/t_{ef}\right)^2 - 0.015\right]$$

This calculation must be carried out for both the minor axis relating to t and the major axis relating to b.

Consider the y-y axis:
$$SR = \frac{2500}{215} = 11.6$$

$$e_x = e_{yy} = 6.0 \text{ mm} = \frac{6.0}{215}t = 0.027t$$

$$e_a = t\left[\frac{1}{2400}(11.6)^2 - 0.015\right] = 0.041t$$

$$e_t = [(0.6 \times 0.027t) + 0.041t] = 0.057t$$

$$e_m = 0.057t$$

Equation 4

β = $1.1[1 - (2e_m /t)]$ = $1.1[1 - (2 \times 0.057t/t)]$
= 0.97

Consider the x-x axis:
$$SR = \frac{5000}{440} = 11.4$$

$$e_x = e_{xx} = 30.0 \text{ mm} = \frac{30.0}{440}b = 0.07t$$

Contract : Column Job Ref. No. : Example 2.13	Calcs. by : W.McK.
Part of Structure : Eccentrically Loaded Column	Checked by :
Calc. Sheet No. : 4 of 5	Date :

References	Calculations	Output
	$e_a = b\left[\dfrac{1}{2400}(11.4)^2 - 0.015\right] = 0.039b$	
	$e_t = [(0.6 \times 0.07b) + 0.039b] = 0.081b$	
	$e_m = 0.081b$	
Equation 4	$\beta = 1.1[1 - (2e_m/b)] = 1.1[1 - (2 \times 0.081b/b)]$ $= 0.92$	
	The critical value of $\beta = 0.92$ **(Note: the smaller value)**	
	$P \leq \dfrac{\beta\,b\,t\,f_k}{\gamma_m} = \dfrac{0.92 \times 440 \times 215 \times 7.9}{3.1 \times 10^3} = 221.8$ kN	$P \leq 221.8$ kN
	Case (iv): $e_{xx} = 40.0$ mm $> 0.05b$ $e_{yy} = 20.0$ mm $> 0.05t$	
	The value for β must be determined using Appendix B:	
	Consider the y-y axis: $SR = \dfrac{2500}{215} = 11.6$	
	$e_x = e_{yy} = 20.0$ mm $= \dfrac{20.0}{215}t = 0.093t$	
	$e_a = t\left[\dfrac{1}{2400}(11.6)^2 - 0.015\right] = 0.041t$	
	$e_t = [(0.6 \times 0.093t) + 0.041t] = 0.097t$	
	$e_m = 0.097t$	
Equation 4	$\beta = 1.1[1 - (2e_m/t)] = 1.1[1 - (2 \times 0.097t/t)]$ $= 0.89$	
	Consider the x-x axis: $SR = \dfrac{5000}{440} = 11.4$	

References	Calculations	Output

Contract : Column   **Job Ref. No. :** Example 2.13

Part of Structure :   Eccentrically Loaded Column

Calc. Sheet No. : 5 of 5

Calcs. by : W.McK.

Checked by :

Date :

References	Calculations	Output
	$e_x = e_{xx} = 40.0 \text{ mm} = \dfrac{40.0}{440}b = 0.091t$	
	$e_a = b\left[\dfrac{1}{2400}(11.4)^2 - 0.015\right] = 0.039b$	
	$e_t = [(0.6 \times 0.091b) + 0.039b] = 0.094b$	
	$e_m = 0.094b$	
Equation 4	$\beta = 1.1[1 - (2e_m/b)] = 1.1[1 - (2 \times 0.094b/b)]$ $\quad = 0.89$	
	The critical value of $\beta = 0.89$	
	$P \le \dfrac{\beta bt f_k}{\gamma_m} = \dfrac{0.89 \times 440 \times 215 \times 7.9}{3.1 \times 10^3} = 214.6 \text{ kN}$	$P \le 214.6 \text{ kN}$

2.21   Example 2.14 Bridge Truss Support Columns

A series of light bridge girders are supported by natural stone masonry columns as shown in Figure 2.65(a). The columns are constructed from blocks with a ratio of height to least horizontal dimension of 2.5. Using the design data given:

   a)   determine the minimum compressive strength of unit required assuming mortar type (iii) is to be used,

   b)   determine a suitable characteristic strength/mortar combination if the piers are constructed from standard format brick as indicated in Figure 2.65(b) instead of the stone blocks.

Design data:

Ultimate load applied on bridge deck (including self-weight)	9.0 kN/m^2
Category of manufacturing control	special
Category of construction control	special
Characteristic self-weight of stone blocks	22.0 kN/m^3
Characteristic self-weight of brickwork	18.0 kN/m^3

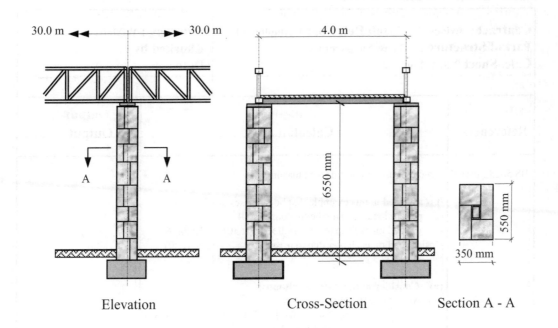

Figure 2.65 (a)

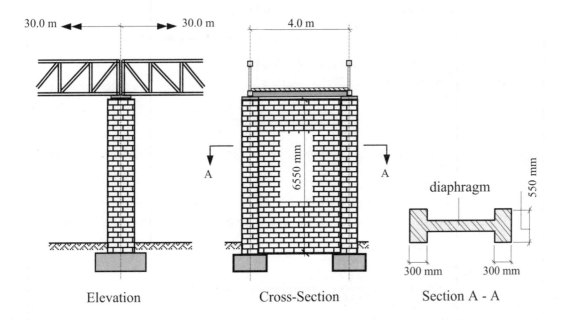

Figure 2.65 (b)

2.21.1 Solution to Example 2.14

References	Calculations	Output
Contract : Bridge **Job Ref. No. :** Example 2.14 **Calcs. by :** W.McK. **Part of Structure :** Truss Support Pier **Checked by :** **Calc. Sheet No. : 1** of **2** **Date :**		

References	Calculations	Output
BS 5628 : Part 1	Structural use of unreinforced masonry	
	It is assumed in this example that both trusses are fully loaded, (i.e. pattern loading is not being considered).	
	Design load on each span $= (9.0 \times 4.0 \times 30) = 1080$ kN	
	Design load on each pier due to trusses $= (0.5 \times 1080)$ $= 540$ kN	
	(a) Consider natural stone columns:	
	Self-weight due to $0.4h$ of column: $= 1.4 \times [(0.4 \times 6.5) \times (22.0 \times 0.55 \times 0.35)]$ $= 15.4$ kN	
	Total design load on the column $= (540 + 15.4) = 555.4$ kN	
Clause 32.2.2	Design vertical load resistance $= \dfrac{\beta\, b\, t\, f_k}{\gamma_m} \geq 555.4$ kN	
Clause 23.1.1	Small plan area: Column area $= (0.35 \times 0.55) = 0.193$ m^2 < 0.2 m^2 Use the multiplying factor $= (0.7 + 1.5A)$ $= [0.7 + (1.5 \times 0.193)] = 0.99$	
Table 4	$\gamma_m = 2.5$	
Clause 28	Slenderness ratio ≤ 27 **minor axis:** $h_{ef} = 6500$ mm, $\qquad\qquad\qquad t_{ef} = 350$ mm $SR = \dfrac{6500}{350} = 18.6; \qquad\qquad e_x \leq 0.05t$	
Table 7	Using interpolation $\beta = 0.75$	
	major axis: $h_{ef} = (2 \times 6500) = 13000$ mm, $\qquad b_{ef} = 550$ mm $SR = \dfrac{13000}{550} = 23.6; \qquad\qquad e_x \leq 0.05t$	
Table 7	Using interpolation $\beta = 0.56$	
	$f_k \geq \dfrac{\left(\gamma_m \times 555.4 \times 10^3\right)}{\left(\beta \times b \times t \times modification\ factor\right)}$ $= \dfrac{\left(2.5 \times 555.4 \times 10^3\right)}{\left(0.56 \times 550 \times 350 \times 0.99\right)} = 13.0$ N/mm^2	

Contract : Bridge Job Ref. No. : Example 2.14	Calcs. by : W.McK.
Part of Structure : Truss Support Pier Calc. Sheet No. : 2 of 2	Checked by : Date :

References	Calculations	Output
Clause 23.1.8	Natural stone masonry should be designed on the basis of solid concrete blocks of an equivalent compressive strength i.e. use Table 2(d) to determine the required unit strength.	**Adopt natural stone blocks with unit strength ≥ 35 N/mm^2 and mortar type (iii)**
Table 2(d)	Assuming mortar designation (iii); aspect ratio $=$ 2.5 i.e. between 2.0 and 4.0 unit strength of 35 N/mm^2 gives f_k > 17 N/mm^2	

	(b) Consider standard format brick piers:
	 Self-weight due to $0.4h$ of pier:
	                $=$ $1.4 \times [(0.4 \times 6.5) \times (18.0 \times 0.55 \times 0.3)]$
	                $=$ 10.8 kN
	 Total design load on the pier $=$ $(540 + 10.8)$ $=$ 550.8 kN	

Clause 32.2.2	Design vertical load resistance $= \dfrac{\beta \, b \, t \, f_k}{\gamma_m}$ $\geq$ 550.8 kN	
Clause 23.1.1	Small plan area: Pier area $=$ $(0.3 \times 0.55) =$ 0.165 m^2 < 0.2 m^2 Use the multiplying factor $=$ $(0.7 + 1.5A)$ $=$ $[0.7 + (1.5 \times 0.165)]$ $= 0.95$	
Table 4	γ_m $=$ 2.5	
Clause 28	Slenderness ratio $\leq$ 27 h_{ef} $=$ 6500 mm, t_{ef} $=$ 550 mm **Note:** the axis of buckling in this case relates to the 550 mm dimension since the wall between the piers provides restraint against the minor axis buckling. SR $= \dfrac{6500}{550}$ $=$ 11.8; e_x $\leq$ $0.05t$	
Table 7	Using interpolation β $=$ 0.93 f_k $\geq$ $\dfrac{\left(\gamma_m \times 550.8 \times 10^3\right)}{\left(\beta \times b \times t \times modification\ factor\right)}$ $= \dfrac{\left(2.5 \times 550.8 \times 10^3\right)}{\left(0.93 \times 550 \times 300 \times 0.95\right)}$ $=$ 9.4 N/mm^2	
Table 2(a)	Assuming mortar designation (iii); unit strength of 50 N/mm^2 gives f_k > 10.6 N/mm^2	**Adopt standard format bricks with unit strength ≥ 50 N/mm^2 and mortar type (iii)**

2.22   Example 2.15   Column Between Adjacent Openings

A column in a 215 mm thick, single leaf masonry wall, is created by two adjacent openings as shown in Figure 2.66. Using the design data given check the suitability of the brick/mortar combination used.

Design data:

Characteristic dead load due to concrete slab, (including self-weight)   10.0 kN/m
Category of manufacturing control   normal
Category of construction control   normal
Characteristic self-weight of brickwork   18.0 kN/m^3
Unit strength of standard format bricks   20.0 N/mm^2
Mortar designation   Type (ii)

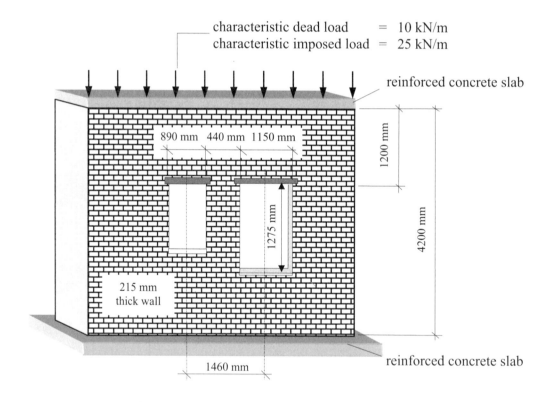

Figure 2.66

2.22.1 Solution to Example 2.15

References	Calculations	Output
	Contract : Wall **Job Ref. No. : Example 2.15** **Part of Structure :** Column Between Adjacent Openings **Calc. Sheet No. : 1 of 1**	**Calcs. by : W.McK.** **Checked by :** **Date :**

References	Calculations	Output
BS 5628 : Part 1	Structural use of unreinforced masonry Width of load supported by column: $\quad = [440 + 0.5(890 + 1150)] \quad = \quad 1.46 \text{ m}$ Design load from slab due to dead load: $\quad = 1.4 \times (10.0 \times 1.46)$ $\qquad\qquad\qquad\qquad\qquad\qquad\qquad = 20.4 \text{ kN}$ Design load from slab due to imposed load:$= 1.6 \times (25.0 \times 1.46)$ $\qquad\qquad\qquad\qquad\qquad\qquad\qquad = 58.4 \text{ kN}$ Design load due to self-weight of $(1.2 \text{ m} + 0.4h)$ of brickwork: $1.4 \times (18.0 \times 0.215 \times 1.2 \times 1.46) +$ $1.4 \times (18.0 \times 0.4 \times 3.0 \times 0.215 \times 0.44) \quad = \quad 12.3 \text{ kN}$ Total design load on column $\quad = \quad (20.4 + 58.4 + 12.3)$ $\qquad\qquad\qquad\qquad\qquad\qquad = \quad 91.1 \text{ kN}$ The concrete slabs at the top and bottom of the wall provide enhanced lateral resistance to the column ends. **Note:** Care must be taken when assessing the extent to which lateral support is provided by the floor/roof construction and the effective height evaluated accordingly.	
Clause 28.3.1.3(a)	$h_{ef} = [(0.75 \times 4200) + (0.25 \times 1275)] \quad = \quad 3469 \text{ mm}$	
	$e_x = \dfrac{(78.8 \times 0.167t)}{91.1} = 0.14t$	
Clause 28	$SR = \dfrac{3469}{215} = 16.13 \quad < 27$	
Table 7	$\beta = 0.72$	
Table 4	$\gamma_m = 3.5$	
Clause 23.1.1	Plan area $= (0.44 \times 0.215)$ $\qquad\qquad = 0.095 \text{ m}^2 \quad < \quad 0.2 \text{ m}^2$ Modification factor $\quad = [0.7 + (1.5 \times 0.095)]$ $\qquad\qquad\qquad\qquad\quad = 0.84$	
Table 2(a)	unit strength $= 20 \text{ N/mm}^2$; mortar type $=$ (ii) $f_k = 6.4 \text{ N/mm}^2$	
Clause 32.2.2	Design load resistance $= \dfrac{\beta \, b \, t \, f_k}{\gamma_m}$ $\qquad = \dfrac{(0.72 \times 215 \times 440 \times 6.4 \times 0.84)}{(3.5 \times 10^3)} = 104.6 \text{ kN} > 91.1 \text{ kN}$	**Unit strength of 20 N/mm² with mortar type (ii) is suitable**

2.23   Review Problems

2.1   Identify eight factors which influence the axial loadbearing capacity of a
       masonry wall.
       (see section 2.1)

2.2   Determine the compressive strength (f_k), of masonry constructed from
       standard format bricks with a unit compressive strength of 35 N/mm^2 and
       a mortar designation Type (iii).
       (see Figure 2.3 : f_k = 8.5 N/mm^2)

2.3   Explain why careful consideration is necessary when selecting a
       combination of structural unit and mortar designation.
       (see section 2.1.1.2)

2.4   Explain the terms '*normal*' and '*special*' control in relation to
       manufacturing of structural units and construction of masonry.
       (see section 2.1.2)

2.5   Explain the reason for the '*small plan area*' factor used in design.
       (see section 2.1.3)

2.6   Explain the reason for the '*narrow brick wall*' factor used in design.
       (see section 2.1.4)

2.7   Define a '*column*' element as used in BS 5628 : Part 1 : 1992.
       (see section 2.1.5.1)

2.8   Describe the difference between simple and enhanced supports.
       (see sections 2.1.5.3 to 2.1.5.6)

2.9   Explain the purpose of the capacity reduction factor 'β'.
       (see section 2.1.7)

2.10   Determine the capacity reduction factor for a wall with a slenderness ratio
       equal to '16' and an eccentricity at the top of the wall equal to '$0.15t$' .
       (see Figure 2.25 : Using interpolation $\beta \approx 0.70$)

2.11   Explain why values of slenderness ratio less than '6' are not included in
       the capacity reduction factor table.
       (see section 2.1.7)

2.12   Explain the reasons for the minimum and maximum specified width of the
       cavity in cavity wall construction.
       (see section 2.7)

2.13 Explain the purpose of the ties in a cavity wall.
(see section 2.7)

2.14 Distinguish between a '*collar-jointed*' wall and a '*grouted cavity*' wall.
(see sections 2.10 and 2.11)

3. Laterally Loaded Walls

Objective: *'To illustrate the requirements for the limit-state design of laterally loaded walls.'*

3.1   Introduction

In many instances for example, cladding panels, masonry must resist forces induced by lateral loading such as wind pressure. The geometric dimensions and support conditions of such panels frequently result in two-way bending as shown in Figure 3.1.

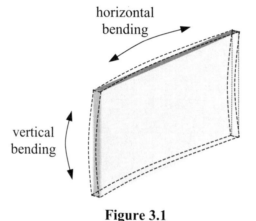

Figure 3.1

Masonry is a non-isotropic material resulting in flexural strengths and modes of failure which are different in the vertical and horizontal directions. The failure mode due to simple vertical bending occurs with cracking developing along the bed joints as shown in Figure 3.2(a) and in simple horizontal bending with cracking developing through the vertical joints as shown in Figure 3.2(b).

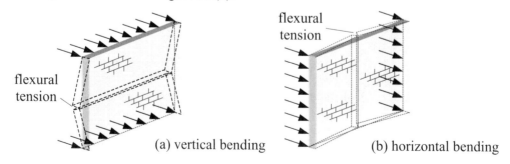

Figure 3.2

In BS 5628 : Part 1 : 1992 Table 3, the characteristic flexural strength of masonry (f_{kx}) is given relative to both individual failure modes as shown in Figure 3.3.

Table 3. Characteristic flexural strength of masonry, f_{kx} in N/mm^2

	$f_{kx\ par}$ 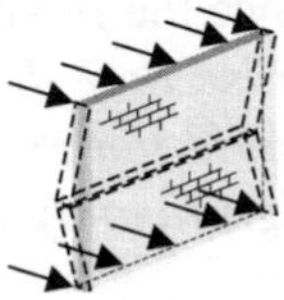			$f_{kx\ perp}$ 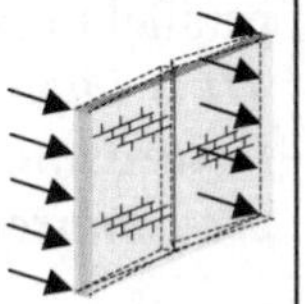		
Mortar Designation	(i)	(ii) & (iii)	(iv)	(i)	(ii) & (iii)	(iv)
Clay bricks having a water absorption						
less than 7%	0.7	0.5	0.40	2.0	1.5	1.2
between 7% and 12%	0.5	0.4	0.35	1.5	1.1	1.0
over 12 %	0.4	0.3	0.25	1.1	0.9	0.8
Calcium silicate bricks	0.3		0.2	0.9		0.6
Concrete bricks	0.3		0.2	0.9		0.6
Concrete blocks (solid or hollow) of compressive strength in N/mm^2						
2.8				0.40		0.4
3.5 }used in walls of thickness * up to 100 mm	0.25		0.2	0.45		0.4
7.0				0.60		0.5
2.8				0.25		0.2
3.5 }used in walls of thickness * up to 250 mm	0.15		0.1	0.25		0.2
7.0				0.35		0.3
10.5				0.75		0.6
14.0 }used in walls of any thickness *	0.25		0.2	0.90 **		0.7 **
and over						

* The thickness should be taken to be the thickness of the wall, for a single-leaf wall, or the thickness of the leaf, for a cavity wall.

** When used with flexural strength in the parallel direction, assume the orthogonal ratio $\mu = 0.3$.

Figure 3.3

As indicated in Clause 24.2 of the code, linear interpolation between the values of f_{kx} is permitted for:

    (a) concrete block walls of thickness between 100 mm and 250 mm,
    (b) concrete blocks of compressive strength between 2.8 N/mm^2 and 7.0 N/mm^2 in a wall of given thickness.

Since the development of flexural tension is clearly an important factor in the flexural strength of masonry any loading which tends to reduce this will enhance the strength of a panel. In loadbearing walls such as the lower levels of multi-storey buildings the precompression caused by the dead load from above will increase the resistance to failure parallel to the bed joints since the flexural tension is reduced by the axial compression.

(It is important to recognise that the increased compressive stresses due to combined flexure and axial load should not exceed their specified limits and in addition, that the pre-compression will be significantly lower at the upper levels of the building.)

The ratio of the flexural strength due to vertical bending to that due to horizontal bending is known as the '**orthogonal ratio**', i.e.:

$$\mu = \frac{f_{kx\,par}}{f_{kx\,perp}}$$

where:

$f_{kx\,par}$ is the characteristic flexural strength parallel to the bed joints

$f_{kx\,perp}$ is the characteristic strength perpendicular to the bed joints

As indicated in Clause 36.4.2 of the code the value of the orthogonal ratio can be modified to reflect the enhanced strength due to pre-compression by increasing the value of $f_{kx\,par}$ accordingly, i.e.:

$$\text{Modified orthogonal ratio} = \frac{\left(\dfrac{f_{kx\,par}}{\gamma_m} + g_d\right)}{\dfrac{f_{kx\,perp}}{\gamma_m}} = \frac{\left(f_{kx\,par} + \gamma_m g_d\right)}{f_{kx\,perp}}$$

where:

$f_{kx\,par}$, $f_{kx\,perp}$ and γ_m are as before

g_d is the design vertical load/unit area. This should be evaluated assuming $\gamma_f = 0.9$ as indicated in Clause 22 of the code.

Wall panels with a high '*height to length*' ratio and with one vertical edge unsupported are particularly sensitive to failure parallel to the bed joints and it is often advantageous to utilise any existing pre-compression to enhance the flexural strength.

3.1.1 *Design Strength of Panels* (Clause 36.4 and Table 9)

The design of panels as given in Clause 36.4 of the code requires that the design moment of resistance is equal to or greater than the calculated design moment due to the applied loads as determined using the bending moment coefficients from Table 9, i.e.

(i)   Consider failure perpendicular to the plane of the bed joints:

Moment of resistance of the panel $= \dfrac{f_{kx\,perp}}{\gamma_m} Z$

Moment due to applied loads $= \alpha W_k \gamma L^2$ per unit height

$$\frac{f_{kx\,perp}}{\gamma_m} Z \geq \alpha W_k \gamma L^2 \qquad (1)$$

(ii)   Consider failure parallel to the plane of the bed joints:

Moment of resistance of the panel $= \dfrac{f_{kx\,par}}{\gamma_m} Z$

Moment due to applied loads $= \mu\alpha W_k \gamma L^2$ per unit height

$$\frac{f_{kx\,par}}{\gamma_m} Z \geq \mu\alpha W_k \gamma L^2 \qquad (2)$$

where:
α is the bending moment coefficient taken from Table 9
γ_f is the partial safety factor for loads (Clause 22)
μ is the orthogonal ratio
L is the length of the panel
W_k is the characteristic wind load/unit area
Z is the elastic section modulus (**Note:** based on the **net** area for hollow blocks)
$f_{kx\,perp}$, $f_{kx\,par}$, and γ_m are as before.

Clause 36.4.3 of the code indicates that the elastic section modulus of a wall including piers should be calculated assuming an outstanding length of flange from the face of the pier should be taken as indicated in Figure 3.4. In no case should the full flange width be greater than the distance between the centre lines of the piers.

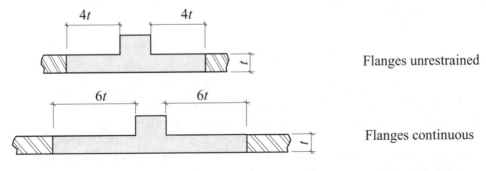

Figure 3.4

The calculation to determine the required flexural strength can be carried out using either equation (1) or equation (2); it is not necessary to evaluate both since the effect of the orthogonal ratio is included in the bending moment coefficients given in Table 9, i.e.

In equation (2)

$$\frac{f_{kx\,par}}{\gamma_m} = \frac{\mu\,f_{kx\,perp}}{\gamma_m} = \mu \times (\alpha W_k \gamma L^2 \text{ per unit height}) = \mu \alpha W_k \gamma L^2 \text{ per unit height}$$

The values of the bending moment coefficients given in Table 9 are dependent on:

    i)   the edge restraint conditions of panels,
    ii) the orthogonal ratio (μ) and
    iii) the aspect ratio h/L where h is the height of the panel and L is the length of the panel.

The edge restraint conditions are summarised in Figure 3.5.

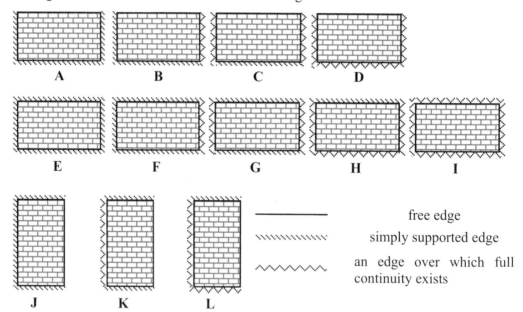

Figure 3.5

The values of the orthogonal ratio vary from 0.3 to 1.0 and the values of the aspect ratio vary from 0.3 to 1.75. Linear interpolation of μ and h/L is permitted. When the dimensions of a wall are outside the range of h/L given in Table 9 it will usually be sufficient to calculate the moments on the basis of a simple span e.g. a panel of type A having h/L less than 0.3 will tend to act as a free standing wall, whilst the same panel having h/L greater than 1.75 will tend to span horizontally.

3.1.2 Edge Support Conditions and Continuity (Clause 36.2)

The lateral resistance of masonry panels is dependent on the degree of rotational and lateral restraint at the edges of the panel and/or the continuity past a support such as a pier or a column. In most cases, with the exception of a free edge, unless masonry is fully bonded into return walls or is in intimate and permanent contact with a roof or floor, a simple support should be assumed.

In BS 5628 : Part 3 guidance is given to assess fixed support conditions for both single-leaf solid walls and cavity walls as shown in Figures 3.6 and 3.7. In all cases a wall should be adequately connected to its support and all supports should be sufficiently strong and rigid to carry the transmitted load.

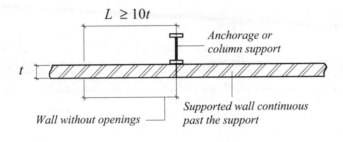

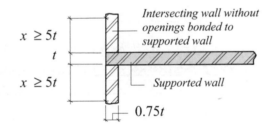

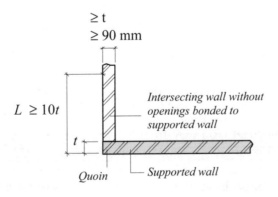

Figure 3.6 Fixed support conditions for solid walls

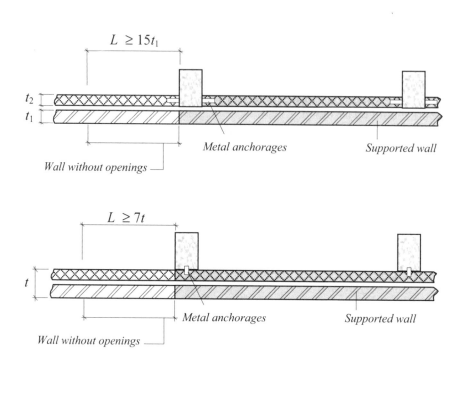

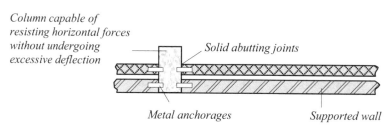

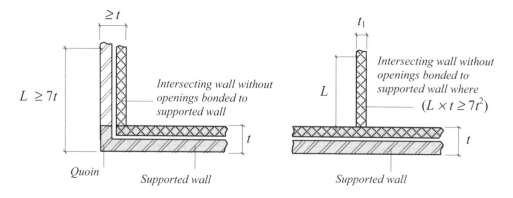

Figure 3.7 Fixed support conditions for cavity walls

In addition to the Part 3 guidelines, BS 5628 : Part 1 : 1992, Figures 7 and 8 provide examples of typical support conditions and continuity over supports as shown in Figures 3.8 and 3.9. The connection to a support may be in the form of ties or by shear resistance of the masonry taking into account the damp proof course, if it exists. Values of characteristic tensile and shear strength which should be used for various types of wall ties used in panel supports are given in Part 1, Table 8 of the code. Where it is necessary to transmit compression, provided that any gap between the wall and the supporting structure is not greater than 75 mm, the values for tension given in Table 8 may be used for ties other than butterfly or double triangle types. As indicated in Clause 36.2, in the case of these types of tie which conform to BS 1243 the characteristic strengths for compression as given in Table 3.1 should be used and '....*In the case of cavity construction, continuity may be assumed even if only one leaf is continuously bonded over or past a support, provided that the cavity wall has ties in accordance with table 6. Where the leaves are of different thicknesses the thicker leaf is to be the continuous leaf. The load to be transmitted from a panel to its support may be taken by ties to one leaf only, provided that there is adequate connection between the two leaves, particularly at the edges of the panels.*'

Characteristic compressive resistance for butterfly and double triangle ties		
Mortar types (i), (ii), (iii) and (iv)	cavity ≤ 75 mm	75 < cavity ≤ 100 mm
Double Triangle Tie	1.25 kN	0.65 kN
Wire Butterfly Tie	0.5 kN	0.35 kN
Note: see Clauses 29.1.4 and 29.1.5 for selection and distribution of ties		

Table 3.1

Metal ties to columns.
Simple support – direct force restraint restricted to values given in Table 8.

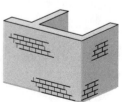

Bonded return walls.
Restrained support – direct force and moment restraint limited by tensile strength of masonry as given in **Clause 24.1**.
(see Chapter 1 Section 1.4.3)

Figure 3.8   Vertical support conditions

Metal ties to columns or unbonded return walls.
Shear and possibly moment restraint.
Shear limited to values given in Table 8.

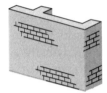

Bonded to piers.
Intermediate pier – direct force and moment
restraint limited by tensile strength of
masonry as given in **Clause 24.1**.
End pier Simple support – direct force
restraint restricted to values given in Table 8.

Figure 3.8 (continued)   Vertical support conditions

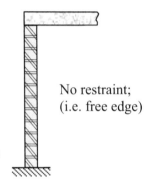

Shear and moment restraint;
limited by flexural and shear
strength of brickwork

Shear and moment restraint;
limited by damp-proof course material
(see **Clauses 24.1** and **33**) and by
vertical load (see **Clause 36.4.2**) of the
code

(a) In-situ floor slab cast on to
     wall span parallel to wall

No restraint;
(i.e. free edge)

(c) Wall built up to but **not** pinned to
     the wall above

Shear and moment restraint;
simple support or moment
restraint at the discretion of
the designer

(b) (1) Precast units spanning parallel to
          the wall
     (2) Walls solidly pinned up to the
          structure above

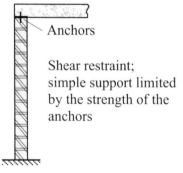

Anchors

Shear restraint;
simple support limited
by the strength of the
anchors

(d) Similar to (c) above with
     suitable anchors

Figure 3.9   Horizontal support conditions

3.1.3 Limiting Dimensions (Clause 36.3)

The limiting dimensions of laterally loaded walls and free-standing walls as set out in Clause 36.3 are indicated in Figure 3.10.

(a) *Panel supported on three edges*

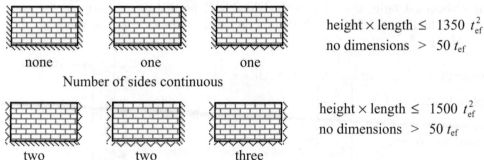

none   one   one

Number of sides continuous

$$\text{height} \times \text{length} \leq 1350\ t_{ef}^2$$
$$\text{no dimensions} > 50\ t_{ef}$$

two   two   three

Number of sides continuous

$$\text{height} \times \text{length} \leq 1500\ t_{ef}^2$$
$$\text{no dimensions} > 50\ t_{ef}$$

(b) *Panel supported on four edges*

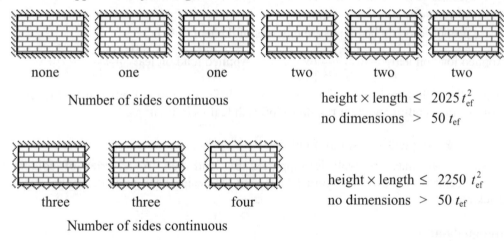

none   one   one   two   two   two

Number of sides continuous

$$\text{height} \times \text{length} \leq 2025\ t_{ef}^2$$
$$\text{no dimensions} > 50\ t_{ef}$$

three   three   four

Number of sides continuous

$$\text{height} \times \text{length} \leq 2250\ t_{ef}^2$$
$$\text{no dimensions} > 50\ t_{ef}$$

(c) *Panel supported top and bottom*

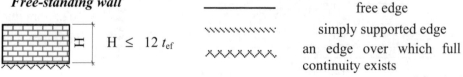

$$H \leq 40\ t_{ef} \qquad\qquad H \leq 40\ t_{ef}$$

(d) *Free-standing wall*

$$H \leq 12\ t_{ef}$$

——————   free edge

〰〰〰〰〰   simply supported edge

⋏⋏⋏⋏⋏⋏   an edge over which full continuity exists

Figure 3.10

3.1.4   Design Lateral Strength of Cavity Walls   (Clause 36.4.5)

The design lateral strength of cavity walls is normally assumed to be equal to the sum of that for each of the two leafs. This is based on the assumption that vertical twist ties, ties with equivalent strength or double-triangle/wire butterfly ties as given in Table 3.1 (see Clause 36.2) are used. Since both leaves are assumed to deflect together, assuming the same orthogonal ratio, the total applied load will be distributed between the walls in proportion to their stiffnesses (**Note:** the relative stiffness ' I ' is adequate for this purpose), e.g.

Assume the total applied lateral load $= W$

Applied load on leaf_1 $= \left[\dfrac{I_1}{I_{\text{total}}} \right] W$

Applied load on leaf_2 $= \left[\dfrac{I_2}{I_{\text{total}}} \right] W$

where:

I_1 is the second moment of area of leaf_1 [i.e. $(t_1^3/12)$/unit length]

I_2 is the second moment of area of leaf_2 [i.e. $(t_2^3/12)$/unit length]

I_{total} $= (I_1 + I_2)$

These proportions can be expressed in terms of the leaf thicknesses:

Applied load on leaf_1 $= \left[\dfrac{t_1^3}{t_1^3 + t_2^3} \right] W$;   Applied load on leaf_2 $= \left[\dfrac{t_2^3}{t_1^3 + t_2^3} \right] W$

Where the orthogonal ratio of the two leaves differ the proportions can be expressed in terms of the bending moment resistance of each leaf as in Example 3.2.

3.2   Example 3.1   Single-Leaf Wall

A single-leaf, masonry wall is supported by, and tied to, a series of columns at 4.25 m centres as shown in Figure 3.11. Using the design data given determine a suitable brick/mortar combination.

Design data:

Characteristic wind load (W_k)   1.2 kN/m^2
Category of manufacturing control   normal
Category of construction control   normal

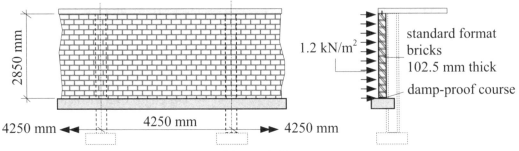

Figure 3.11

3.2.1   Solution to Example 3.1

References	Calculations	Output
	Contract : Workshop **Job Ref. No. :** Example 3.1 **Calcs. by :** W.McK. **Part of Structure :** Wall Panel **Checked by :** **Calc. Sheet No. : 1** of **3** **Date :**	

References	Calculations	Output
BS 5628 : Part 1	Structural use of unreinforced masonry	
Clause 36.2 Figures 7 and 8	Assume simple supports at the top and bottom of the wall. The continuity of the wall over the columns provides fixity to the vertical edges of each panel. Panel considered for design:	

References	Calculations	Output
Clause 36.3 (Figure 3.10)	Limiting dimensions: The panel is supported on four edges with two sides continuous. height $\times$ length $\leq$ 2025 t_{ef}^2 and No dimension $> 50 t_{ef}$ t_{ef} = 102.5 mm, $\therefore$ 2025 t_{ef}^2 = 21.28 $\times 10^6$ mm^2 $\qquad\qquad\qquad\qquad$ 50t_{ef} = 5125 mm height $\times$ length = (2850 $\times$ 4250) = 12.11 $\times 10^6$ mm^2 $\qquad\qquad\qquad\qquad$ < 2025 t_{ef}^2 height = 2850 mm < 50t_{ef} length = 4250 mm < 50t_{ef}	**Limiting dimensions** **are satisfied**
Table 9 (Figure 3.5)	The panel considered corresponds to Type G. The orthogonal ratio (μ) is required to obtain a value for the bending moment coefficient, since this is dependent on the type of brick and mortar used, assume a value equal to 0.35 and check at the end of the calculation. $\mu = 0.35$; $h/L = (2850/4250)$ = 0.67	

Contract : Workshop Job Ref. No. : Example 3.1	Calcs. by : W.McK.
Part of Structure : Wall Panel	Checked by :
Calc. Sheet No. : 2 of 3	Date :

References	Calculations	Output
Table 9	(Extract relating to panel type G)	

(Extract relating to panel type G)

μ			h/L		
	0.3	**0.5**	**0.75**	1.00	1.25
1.00	0.007	0.014	0.022	0.028	0.033
0.90	0.008	0.015	0.023	0.029	0.034
0.80	0.008	0.016	0.024	0.031	0.035
0.70	0.009	0.017	0.026	0.032	0.037
0.60	0.010	0.019	0.028	0.034	0.038
0.50	0.011	0.021	0.030	0.036	0.04
0.40	0.013	0.023	0.032	0.038	0.042
0.35	0.014	0.025	0.033	0.039	0.043
0.30	0.016	0.026	0.035	0.041	0.044

Interpolation (for $h/L = 0.67$) gives $\alpha = 0.03$

Clause 36.4.2   Consider bending perpendicular to the bed joints.

Design bending moment $= (\alpha\, W_k \gamma_f L^2)/$ metre height.

Clause 22 $\gamma_f = 1.4$

Note:
'*In the particular case of freestanding walls and laterally loaded wall panels, whose removal would in no way affect the stability of the remaining structure, γ_f applied on the wind load may be taken as 1.2*'

Design bending moment $= (0.03 \times 1.2 \times 1.4 \times 4.25^2)$
$= 0.91$ kNm /metre height

Clause 36.4.3   Design moment of resistance $= \left(\dfrac{f_{kx\,perp}}{\gamma_m} Z \right) \geq 0.91 \times 10^6$

Table 4 $\gamma_f = 3.5$

$Z = \dfrac{1000 \times 102.5^2}{6} = (1.751 \times 10^6)\ \text{mm}^3 /$ metre height

$\dfrac{f_{kx\,perp} \times 1.751 \times 10^6}{3.5} \geq 0.91 \times 10^6\ \text{Nmm / metre height}$

$f_{kx\,perp} \geq 1.82\ \text{N/mm}^2$

Contract : Workshop Job Ref. No. : Example 3.1	Calcs. by : W.McK.
Part of Structure : Wall Panel	Checked by :
Calc. Sheet No. : 3 of 3	Date :

References	Calculations	Output
Table 3	(Extract from Table 3)	

Table 3. Characteristic flexural strength of masonry f_{kx}, N/mm^2

		$f_{kx\ par}$			$f_{kx\ perp}$		
Mortar Designation		(i)	(ii) and (iii)	(iv)	**(i)**	(ii) and (iii)	(iv)
Clay bricks with water absorption:	< 7%	0.7	0.5	0.4	2.0	1.5	1.2
	≥ 7% and ≤ 12%	0.5	0.4	0.35	1.5	1.1	1.0
	>12%	0.4	0.3	0.25	1.1	0.9	0.8
Calcium silicate bricks		0.3		0.2	0.9		0.6
Concrete bricks		0.3		0.2	0.9		0.6

Use clay bricks having a water absorption less than 7% with a mortar designation Type (i).

Check the assumed value of the orthogonal ratio μ

$$\mu = \frac{f_{kx\ par}}{f_{kx\ perp}} = \frac{0.7}{2.0} = 0.35$$

In this case the assumed value of μ was 0.35. If the actual value after selecting a suitable combination of bricks and mortar had been lower than the assumed value then the correct bending moment coefficient would be slightly higher. A consequence of this is that the design bending moment would be higher and a check on the revised value for the selected combination would be necessary.

Note:
The reader should repeat this calculation based on the flexural strength parallel to the bed joints.

Output:
Adopt clay bricks with a water absorption of less than 7% with a mortar designation Type (i)

3.3   Example 3.2   Cavity Wall

The side cladding for a steel portal framed building is to be of cavity wall construction as shown in Figure 3.12. At the base the wall is supported on a dropped edge beam which is incorporated in the floor slab and is built off a bituminous damp-proof course. The wall is built up to, but not pinned to, the structure above. Using the design data given and considering a typical internal panel:

(i)   determine a suitable brick/mortar combination assuming both leafs are constructed from calcium silicate bricks,   and

(ii) determine the maximum wind load which can be resisted if the outer leaf is constructed from calcium silicate bricks and the inner leaf is constructed from 90 mm thick hollow concrete blocks with a compressive strength of 7 N/mm². Both leaves are constructed using mortar type (i).

Design data:
Characteristic wind load (W_k) for (i)   0.5 kN/m²
Category of manufacturing/construction control   special/normal

Note: Assume that the wall does not provide lateral bracing to the building and its removal does not affect the stability of the remaining structure.

Vertical twist wall ties complying with Clauses 29.1.4, 29.1.5 and 36.2 are to be used and all cladding panels are to be tied to the steel frame using suitable anchors.

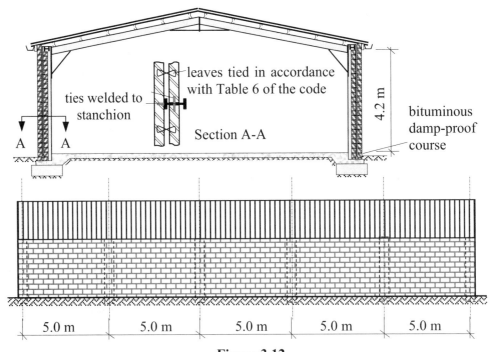

Figure 3.12

3.3.1 Solution to Example 3.2

Contract : Steel Portal Frame Job Ref. No. : Example 3.2	Calcs. by : W.McK.
Part of Structure : Cavity Wall Panel	Checked by :
Calc. Sheet No. : 1 of 7	Date :

References	Calculations	Output
BS 5628 : Part 1	Structural use of unreinforced masonry	
Figures 7 and 8	Assume a free edge at the top, a simple support at the bottom and fixed vertical edges on the panel.	
Clause 36.2	In the case of cavity walls, only one leaf need be continuous, (the thicker of the two), provided wall ties in accordance with Table 6 are used between the two leaves and between each section of the discontinuous leaf and the support.	

Panel considered for design:

(a) **Consider two leaves of calcium silicate bricks**

Clause 36.3 (Figure 3.10)	Limiting dimensions: The panel is supported on three edges with two sides continuous.

$$\text{height} \times \text{length} \leq 1500\, t_{ef}^2 \quad \text{and} \quad \text{No dimension} > 50 t_{ef}$$

Clause 28.4.1	Effective thickness For cavity walls the effective thickness is as indicated in Figure 3 of the code and equal to the greatest of:

(a) $2/3(t_1 + t_2)$ = $2/3(102.5 + 102.5)$ = 136.7 mm or
(b) t_1 = 102.5 mm or
(c) t_2 = 75 mm

$\therefore\ t_{ef}$ = 136.7 mm $\therefore\ 1500\, t_{ef}^2$ = 28.03×10^6 mm^2

$50 t_{ef}$ = 6835 mm

References	Calculations	Output
	Contract : Steel Portal Frame **Job Ref. No. :** Example 3.2 **Calcs. by :** W.McK. **Part of Structure :** Cavity Wall Panel **Checked by :** **Calc. Sheet No. : 2** of **7** **Date :**	

References	Calculations	Output
	height × length = (4200 × 5000) = 21.0×10^6 mm^2 < $1500\, t_{ef}^2$ height = 4200 mm < $50t_{ef}$ length = 5000 mm < $50t_{ef}$	Limiting dimensions are satisfied
Table 9 (Figure 3.5)	The panel considered corresponds to Type C. The orthogonal ratio (μ) is required to obtain a value for the bending moment coefficient, since this is dependent on the type of brick and mortar used, assume a value equal to 0.35 and check at the end of the calculation. $\mu = 0.35$; $h/L = (4200/5000) = 0.84$	
Table 9	(Extract relating to panel type C)	

	h /L				
μ	0.3	0.5	**0.75**	**1.00**	1.25
1.00	0.020	0.028	0.037	0.042	0.045
0.90	0.021	0.029	0.038	0.043	0.046
0.80	0.022	0.031	0.039	0.043	0.047
0.70	0.023	0.032	0.040	0.044	0.048
0.60	0.024	0.034	0.041	0.046	0.049
0.50	0.025	0.035	0.043	0.047	0.050
0.40	0.027	0.038	0.044	0.048	0.051
0.35	0.029	0.0398	0.045	0.049	0.052
0.30	0.030	0.040	0.046	0.050	0.052

References	Calculations	Output
	Interpolation (for $h/L = 0.84$) gives $\alpha = 0.046$	
Clause 36.4.2	Consider bending perpendicular to the bed joints. Design bending moment = $(\alpha\, W_k \gamma_f L^2)$/ metre height.	
Clause 22	**Note:** *'In the particular case of freestanding walls and laterally loaded wall panels, whose removal would in no way affect the stability of the remaining structure, γ_f applied on the wind load may be taken as 1.2'* $\therefore \ \gamma_f = 1.2$ Design bending moment = $(0.046 \times 0.5 \times 1.2 \times 5.0^2)$ = 0.69 kNm /metre height	

References	Calculations	Output

Contract : Steel Portal Frame **Job Ref. No. :** Example 3.2 **Calcs. by :** W.McK.
Part of Structure : Cavity Wall Panel **Checked by :**
Calc. Sheet No. : 3 of 7 **Date :**

Clause 36.4.5 — Since both leaves are the same the bending moment carried by
one leaf $= (0.5 \times 0.69)$ = 0.345 kNm /metre height

Clause 36.4.3 — Design moment of resistance $= \left(\dfrac{f_{kx\ perp}}{\gamma_m} Z \right) \geq 0.345 \times 10^6$

Table 4 — $\gamma_m = 3.1$

$$Z = \frac{1000 \times 102.5^2}{6} = (1.751 \times 10^6) \text{ mm}^3 / \text{metre height}$$

$$\frac{f_{kx\ perp} \times 1.751 \times 10^6}{3.1} \geq 0.345 \times 10^6 \text{ Nmm} / \text{metre height}$$

$$f_{kx\ perp} \geq 0.61 \text{ N/mm}^2$$

Table 3 — (Extract from Table 3)

Table 3. Characteristic flexural strength of masonry f_{kx}, N/mm^2

Mortar Designation	$f_{kx\ par}$			$f_{kx\ perp}$		
	(i)	(ii) and (iii)	(iv)	(i)	(ii) and (iii)	(iv)
Clay bricks with water absorption:						
< 7%	0.7	0.5	0.4	2.0	1.5	1.2
≥ 7% and ≤ 12%	0.5	0.4	0.35	1.5	1.1	1.0
>12%	0.4	0.3	0.25	1.1	0.9	0.8
Calcium silicate bricks	0.3		0.2	0.9		0.6
Concrete bricks	0.3		0.2	0.9		0.6
Concrete blocks (solid or hollow) of compressive strength in N/mm^2						
2.8 used in walls				0.40		0.4
3.5 up to 100 mm	0.25		0.2	0.45		0.4
7.0 thick				0.60		0.5

Contract : Steel Portal Frame Job Ref. No. : Example 3.2	Calcs. by : W.McK.
Part of Structure : Cavity Wall Panel Calc. Sheet No. : 4 of 7	Checked by : Date :

References	Calculations	Output
	Use calcium silicate bricks with a mortar designation Type (iii). ($f_{kx\,perp}$ = 0.9 N/mm²) Check the assumed value of the orthogonal ratio μ $\mu = \dfrac{f_{kx\,par}}{f_{kx\,perp}} = \dfrac{0.3}{0.9} = 0.33$	

Table 9 — Determine the value of α corresponding to $\mu = 0.33$

	h /L				
μ	0.3	0.5	**0.75**	**1.00**	1.25
1.00	0.020	0.028	0.037	0.042	0.045
0.90	0.021	0.029	0.038	0.043	0.046
0.80	0.022	0.031	0.039	0.043	0.047
0.70	0.023	0.032	0.040	0.044	0.048
0.60	0.024	0.034	0.041	0.046	0.049
0.50	0.025	0.035	0.043	0.047	0.050
0.40	0.027	0.038	0.044	0.048	0.051
0.35	0.029	0.0398	0.045	0.049	0.052
0.30	0.030	0.040	0.046	0.050	0.052

$\alpha = 0.047$

Using proportion the required $f_{kx\,perp}$ = $\left(0.61 \times \dfrac{0.047}{0.046}\right)$

= 0.62 N/mm²

< 0.9 N/mm²

Output: Adopt calcium silicate bricks with a mortar designation type (i)

(b) Consider one leaf of calcium silicate bricks and one leaf of 90 mm thick hollow concrete blocks with compressive strength 7N/mm²; both leaves in mortar type (i).

Clause 36.3 / (Figure 3.10)

Limiting dimensions are as before:
The panel is supported on four edges with two sides continuous.

height × length ≤ $1500\,t_{ef}^{2}$ and No dimension > $50t_{ef}$

Clause 28.4.1

Effective thickness
For cavity walls the effective thickness is as indicated in Figure 3 of the code and equal to the greatest of:
(a) $2/3(t_1 + t_2)$ = $2/3(102.5 + 90)$ = 128.3 mm or
(b) t_1 = 102.5 mm or
(c) t_2 = 90 mm

$\therefore\ t_{ef}$ = 128.3 mm $\therefore\ 1500\,t_{ef}^{2}$ = 24.69×10^{6} mm²

$50t_{ef}$ = 6415 mm

References	Calculations	Output

Contract : Steel Portal Frame **Job Ref. No. :** Example 3.2   **Calcs. by : W.McK.**
Part of Structure : Cavity Wall Panel   **Checked by :**
Calc. Sheet No. : 5 of 7   **Date :**

References	Calculations	Output
	height × length = (4200 × 5000) = 21.0×10^6 mm² < $1500 \, t_{ef}^2$	
	height = 4200 mm < $50 t_{ef}$ length = 5000 mm < $50 t_{ef}$	**Limiting dimensions are satisfied**
Clause 36.4.2	Consider bending perpendicular to the bed joints.	
(see section 3.1.4)	Since both leaves have different stiffnesses **and** orthogonal ratios the proportion of the total load carried by each one will be dependent on the bending moment resistance of each one.	
Table 9 (Figure 3.5)	The panel considered corresponds to Type C.	
Table 3 Table 4	***Consider the outer leaf (t_1 = 102.5 mm)*** $f_{kx \, par}$ = 0.3 N/mm² $f_{kx \, perp}$ = 0.9 N/mm² γ_m = 3.1 Z = (1.751×10^6) mm³ / metre height	
Clause 36.4.3	Design moment of resistance = $\left(\dfrac{f_{kx \, perp}}{\gamma_m} Z \right)$ = $\dfrac{0.9 \times 1.751}{3.1}$ = 0.508 kNm/metre height μ = $\dfrac{f_{kx \, par}}{f_{kx \, perp}}$ = $\dfrac{0.3}{0.9}$ = 0.33 $h/L = (4200/5000)$ = 0.84	
Table 9	Interpolation gives α = 0.047	
Clause 22	γ_f = 1.2 as before.	
Clause 36.4.2	Design bending moment = $(\alpha \, W_k \gamma_f L^2)$/ metre height = $(0.047 \times W_k \times 1.2 \times 5.0^2)$ = $(1.41 \times W_k)$ Design moment of resistance ≥ $(1.41 \times W_k)$ $W_k \leq \left(\dfrac{0.508}{1.41} \right)$ = 0.36 kN/m²	
Table 3 Table 4	***Consider the inner leaf (t_2 = 90 mm)*** $f_{kx \, par}$ = 0.25 N/mm² $f_{kx \, perp}$ = 0.6 N/mm² γ_m = 3.1	

Contract : Steel Portal Frame Job Ref. No. : Example 3.2	Calcs. by : W.McK.
Part of Structure : Cavity Wall Panel	Checked by :
Calc. Sheet No. : 7 of 7	Date :

References	Calculations	Output
	$Z = \dfrac{1000 \times 90^2}{6} = (1.35 \times 10^6) \text{ mm}^3 / \text{metre height}$	
Clause 36.4.3	Design moment of resistance $= \left(\dfrac{f_{kx\,perp}}{\gamma_m} Z \right)$	
	$= \dfrac{0.6 \times 1.35}{3.1} = 0.261 \text{ kNm/metre height}$	
	$\mu = \dfrac{f_{kx\,par}}{f_{kx\,perp}} = \dfrac{0.25}{0.6} = 0.41$	
	$h/L = (4200/5000) = 0.84$	
Table 9	Interpolation gives $\alpha = 0.045$	
Clause 22	$\gamma_f = 1.2$ as before.	
Clause 36.4.2	Design bending moment $= (\alpha W_k \gamma_f L^2)/$ metre height $= (0.045 \times W_k \times 1.2 \times 5.0^2)$ $= (1.35 \times W_k)$	
	Design moment of resistance $\geq (1.35 \times W_k)$	
	$W_k \leq \left(\dfrac{0.261}{1.35} \right) = 0.19 \text{ kN/m}^2$	
Clause 36.4.5	Design lateral strength $= \sum$ design strengths of both leaves	**Maximum ultimate**
	Total wind load permitted $= (0.36 + 0.19) = 0.55 \text{ kN/m}^2$	**wind load which can be resisted = 0.55 kN/m^2**

3.4   Example 3.3   Single-Leaf Wall with Pre-Compression

A vertically spanning wall panel in an exhibition centre, which can be considered to be simply supported at the top and the bottom is shown in Figure 3.13. Using the data provided determine the suitability of the panel to resist the wind pressure indicated.

Design data:

Characteristic wind load (W_k)	0.6 kN/m^2
Category of manufacturing control	special
Category of construction control	special
Characteristic vertical load on panel	30 kN/m
Characteristic unit weight of brickwork	18.0 kN/m^3

Masonry is constructed from clay bricks having water absorption between 7% and 12% with mortar designation Type (i).

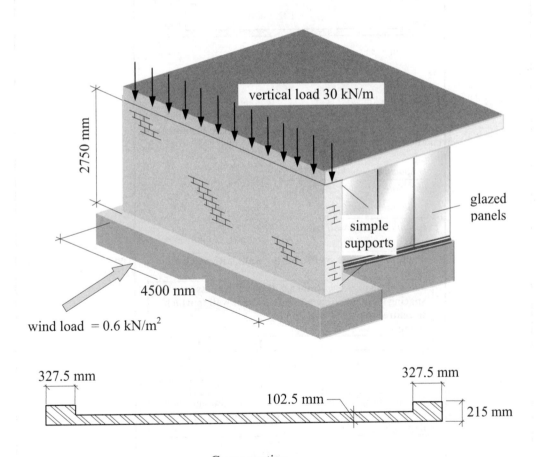

Cross-section

Figure 3.13

3.4.1   Solution to Example 3.3

Contract : Exhibition Centre Job Ref. No. : Example 3.3 Part of Structure : Vertically Spanning Wall Panel Calc. Sheet No. : 1 of 7	Calcs. by : W.McK. Checked by : Date :

References	Calculations	Output
BS 5628 : Part 1	Structural use of unreinforced masonry	
Clause 36.3 (c) (Figure 3.10)	Limiting dimensions: The panel is simply supported at top and bottom. height $\leq$ 40 t_{ef}	

327.5 mm   327.5 mm

102.5 mm

4500 mm   215 mm

| Clause 28.4.1 | Effective thickness For walls stiffened by piers the effective thickness is as indicated in Figure 3 of the code and equal to $(t \times k)$ where k is a stiffness coefficient obtained from Table 5. | |
| Table 5 | $$\frac{pier\ spacing}{pier\ width} = \frac{(4500 - 327.5)}{327.5} = 12.74$$ $$\frac{pier\ thickness\ (t_p)}{actual\ thickness\ of\ wall\ (t)} = \frac{215}{102.5} = 2.1$$ | |

Table 5. Stiffness coefficient for walls stiffened by piers

Ratio of pier spacing (centre to centre) to pier width	Ratio t_p/t of pier thickness to actual thickness of wall to which it is bonded		
	1	2	3
6	1.0	1.4	2.0
10	1.0	1.2	1.4
20	1.0	1.0	1.0

NOTE: Linear interpolation between the values given in table 5 is permissible, but not extrapolation outside the limits given.

Using linear interpolation $k = 1.16$
$t_{ef} = t \times k = (102.5 \times 1.16)$ 118.9 mm
$40\ t_{ef} = (40 \times 118.9) = 4756$ mm

Height = 2750 mm < 40 t_{ef}

Limiting dimensions are satisfied

Contract : Exhibition Centre Job Ref. No. : Example 3.3	Calcs. by : W.McK.
Part of Structure : Vertically Spanning Wall Panel	Checked by :
Calc. Sheet No. : 2 of 7	Date :

References	Calculations	Output
Clause 36.4.2	For walls spanning vertically, the design moment/unit length may be taken as:	
	Design bending moment $= (W_k \gamma_f h^2)/8$	
Clause 22	$\gamma_f = 1.4$	
	Design bending moment $= (0.6 \times 1.4 \times 2.75^2 \times 4.5)/8$ $= 3.57$ kNm	
Clause 36.4.3	Design moment of resistance $= \left(\dfrac{f_{kx\,par}}{\gamma_m} Z \right)$	
Table 4	$\gamma_m = 2.5$	
Table 3	For clay bricks having water absorption between 7% and 12% and mortar designation (i): $f_{kx\,par} = 0.5$ N/mm^2	
Clause 36.4.2	Since the wall panel is also supporting a vertical load, this can be used to enhance the flexural strength of the masonry parallel to the bed joints.	
	Cross-sectional area of the wall: $= [(4.5 \times 0.1025) + (0.3275 \times 0.1125)] = 0.498$ m^2	
	Characteristic self-weight of the upper half of the brickwork: $= (18.0 \times 0.498 \times 0.5 \times 2.75) = 12.32$ kN	
	Total vertical dead load $= [(30 \times 4.25) + 12.32] = 139.82$ kN	
Clause 22	Since this effect is beneficial to the strength use $\gamma_f = 0.9$	
	Axial load considered to enhance $f_{kx\,par}$: $= (0.9 \times 139.82) = 125.8$ kN	
	Design vertical load/unit area $= \dfrac{125.8 \times 10^3}{0.498 \times 10^6}$ $= 0.25$ N/mm^2	
	Enhanced flexural strength $= (f_{kx\,ppar} + \gamma_m g_d)$ $= [0.5 + (2.5 \times 0.25)]$ $= 1.125$ N/mm^2	

Contract : **Exhibition Centre** Job **Ref. No. :** Example 3.3 Part of Structure : **Vertically Spanning Wall Panel** Calc. Sheet No. : **3** of **7**	Calcs. by : **W.McK.** Checked by : Date :

References	Calculations	Output
	Section properties: 327.5 mm 102.5 mm 327.5 mm 215 mm 4500 mm Position of neutral axis: $\bar{y} = \dfrac{[(4500 \times 102.5 \times 51.25) + (2 \times 327.5 \times 112.5 \times 158.75)]}{(498 \times 10^3)}$ $\bar{y} = 70.96$ mm 327.5 x-x axis 327.5 144.04 87.79 70.96 −19.71 4500 $I_{xx} = \left[\left(\dfrac{4500 \times 102.5^3}{12} \right) + \left(4500 \times 102.5 \times 19.71^2 \right) \right] +$ $\qquad 2 \times \left[\left(\dfrac{327.5 \times 112.5^3}{12} \right) + \left(327.5 \times 112.5 \times 87.79^2 \right) \right]$ $\qquad = 1228.6 \times 10^6$ mm^4 $Z_{\text{tension face}} = \dfrac{1228.6 \times 10^6}{144.04} = (8.53 \times 10^6)$ mm^3 Design moment of resistance $= \left(\dfrac{f_{\text{kx par}}}{\gamma_m} Z \right)$ $\qquad = \dfrac{1.125 \times 8.53 \times 10^6}{2.5 \times 10^6}$ $\qquad = 3.84$ kNm ≥ 3.57 kNm	
Clause 36.4.3		**The wall panel is suitable to resist the given wind pressure**
	Note: The wall would not have been suitable had advantage not been taken of the enhanced strength due to the vertical load.	

3.5 Freestanding Walls (Clause 36.5)

Extensive use is made of freestanding walls for landscaping, screening, parapets, internal balustrades, boundary demarcation, security and noise barriers. There are numerous plan arrangements which can be used, some of the more frequently used ones are shown in Figure 3.14.

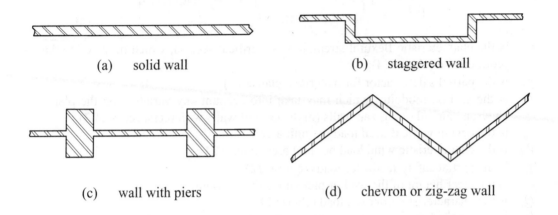

(a)   solid wall   (b)   staggered wall

(c)   wall with piers   (d)   chevron or zig-zag wall

Figure 3.14

The wall configuration used in any particular circumstance will be dependent on a number of criteria such as strength, durability, aesthetics and economy. In many cases it is necessary to resist wind pressure which can be applied on either side of a wall and consequently it may be more efficient to position piers in relation to a wall as shown in Figure 3.14(c), although they need not be symmetrically disposed. The chevron arrangement shown in Figure 3.14(d) is particularly efficient when resisting large bending moments, is inherently stiffer than solid walls of the same thickness as in Figure 3.14(a) and in addition is aesthetically more pleasing.

The configurations indicated in Figure 3.14 are usually economical for wall heights up to approximately 5.0 m. Alternative solutions such as diaphragm wall construction (see Chapter 1, Section 1.3.5), or reinforced brickwork (see Chapter 4) can be considered for higher, more heavily loaded walls.

The design of freestanding walls is governed by the requirements of Clause 36.5 of the code in which it is stated that '*Freestanding walls should be designed as cantilevers springing from the top of the foundation or from the point of horizontal lateral restraint when such restraint is sufficient to resist the horizontal reaction from the wall, except that wall panels between piers may be designed as three-sided or horizontally spanning in accordance with 36.4.3 and 36.4.4 respectively. The piers should then be designed as cantilevers to resist the reaction from the panel. Mortar should not be weaker than designation (iii) (see table 1).*'

In cases where the damp-proof course consists of material which has been proved by tests to permit the joints to transmit tension (see DD 86 : Part 1) or is of bricks complying with the requirements of BS 743 such that flexural tension can be relied upon (e.g. two courses of bricks having absorption of not more than 7%, or two courses of slates fully half-lapped and bedded in mortar) the following relationship must be satisfied:

$$\text{Design moment of resistance} \quad \geq \quad \text{Applied design moment}$$

$$\left[\left(\frac{f_{kx}}{\gamma_m} + g_d\right) \times Z\right] \quad \geq \quad \left[\left(W_k \gamma_f \frac{h^2}{2}\right) + \left(Q_k \gamma_f h_L\right)\right]$$

where:

f_{kx} is the characteristic flexural strength at the critical section, which may be the damp-proof course (clause **24**);

γ_m is the partial safety factor for materials (clause **27**);

Z is the section modulus, which may take into account any variation on the plan, e.g. chevron, curved or zig-zag walls (in the case of walls with piers, see **36.4.3**);

g_d is the design vertical dead load per unit area;

W_k is the characteristic wind load per unit area (clause **21**);

γ_f is the partial safety factor for loads (clause **22**);

h is the clear height of the wall or pier above the restriant;

Q_k is the characteristic imposed load (clause **21**);

h_L is the vertical distance between the point of application of the horizontal load Q_k and the lateral restraint,

as indicated in Clauses **36.5.3** and **36.5.2**.

3.6   Example 3.4   Freestanding Boundary Wall

A freestanding boundary wall in a leisure park is shown in Figure 3.15 (a) with four alternative plan configurations as indicated in Figures 3.15 (b), (c), (d) and (e). Determine the maximum value of wind load (W_k) for each configuration.

Design data:

Category of manufacturing control   special
Category of construction control   normal
Characteristic unit weight of brickwork   18.0 kN/m^3
All walls are constructed from clay bricks having a water absorption over 12% with a mortar designation Type (ii)

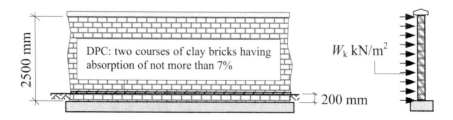

Figure 3.15 (a)

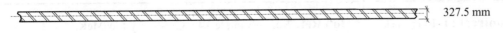

327.5 mm

(b) **solid wall**

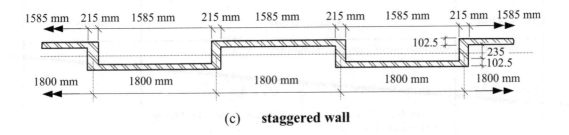

(c) **staggered wall**

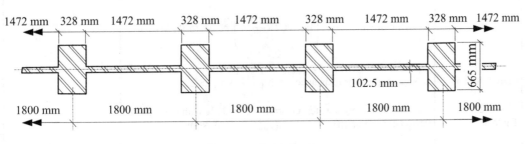

(d) **wall with piers**

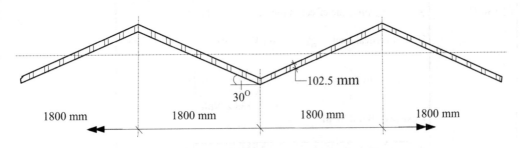

(e) **chevron or zig-zag wall**

Figure 3.15 (continued)

3.6.1   Solution to Example 3.4

Contract : Leisure Park Job Ref. No. : Example 3.4	Calcs. by : W.McK.
Part of Structure : Freestanding Wall	Checked by :
Calc. Sheet No. : 1 of 5	Date :

References	Calculations	Output
BS 5628 : Part 1	Structural use of unreinforced masonry	
Clause 36.5.2	Design moment $= W_k \gamma_f \dfrac{h^2}{L} + Q_k \gamma_f h_L$	
	In this problem there is no additional lateral imposed load being considered and hence $Q_k = 0$	
Clause 22	$\gamma_f = 1.2$ for freestanding walls h is measured from the top of the foundation. $h = 2.5$ m	
	Design moment $= \left(W_k \times 1.2 \times \dfrac{2.5^2}{2} \right) = 3.75\ W_k$ kNm/m	
Clause 36.5.3	Design moment of resistance $= \left(\dfrac{f_{kx}}{\gamma_m} + g_d \right) \times Z$	
Clause 24.0 Table 3	Clay bricks having a water absorption over 12% with a mortar of designation Type (ii) $f_{kx\,par} = 0.3$ N/mm^2	
Clause 27.0 Table 4	$\gamma_m = 3.1$	
	The design vertical load (g_d) is due to the self-weight of the wall. Self-weight $= (18.0 \times 2.5) = 45.0$ kN/m^2 $(= 0.045$ N/mm$^2)$	
Clause 22	$\gamma_f = 0.9$ since the effect of the dead load is beneficial.	
	Design vertical load $g_d = (0.045 \times 0.9) = 0.041$ N/mm^2	
	Design moment of resistance $= \left(\dfrac{0.3}{3.1} + 0.041 \right) \times Z$ $= 0.138\ Z$ Nmm	
	Case (a) – Solid Wall	
	1000 mm 327.5 mm	
	$Z = \dfrac{1000 \times 327.5^2}{6} = 17.87 \times 10^6$ mm^3	

Contract : Leisure Park Job Ref. No. : Example 3.4	Calcs. by : W.McK.
Part of Structure : Freestanding Wall	Checked by :
Calc. Sheet No. : **2** of **5**	Date :

References	Calculations	Output
	Design moment of resistance $= \;(0.146 \times 17.87 \times 10^6)/10^6$ $= \;$ 2.46 kNm/metre run Design moment $\;= 3.75 \, W_k \leq \; 2.46 \;\therefore\; W_k \leq \; 0.66$ kN/m^2 **Case (b) – Staggered Wall** 215 mm 1585 mm 215 mm 102.5 mm 337.5 mm 1800 mm 1800 mm	$W_k \leq \;$ **0.66 kN/m^2**
Clause 36.4.3	The outstanding flange length from the face of a pier is equal to: (6 × thickness of the wall forming the flange) $\;=\; (6 \times 102.5)$ $=\; 615$ mm 440 mm 168.75 mm 102.5 mm 235 mm 102.5 mm 615 mm 615 mm 215 mm $I_{xx} = \left(\dfrac{215 \times 440^3}{12}\right) + 2\left[\dfrac{615 \times 102.5^3}{12} + \left(615 \times 102.5 \times 168.75^2\right)\right]$ $I_{xx} = 5226\,8 \times 10^6$ mm^4 Average $I_{xx} \;=\; \dfrac{5226.8 \times 10^6}{1.8} \;=\; 2903 \times 10^6$ mm^4 Average $Z_{xx} \;=\; \dfrac{2903 \times 10^6}{220} \;=\; 13.2 \times 10^6$ mm^3 Design moment of resistance $=\; (0.146 \times 13.2 \times 10^6)/10^6$ $=\; 1.93$ kNm/metre run Design moment $\;= 3.75 \, W_k \leq \; 1.93 \;\therefore\; W_k \leq \; 0.51$ kN/m^2	

Contract : Leisure Park Job Ref. No. : Example 3.4	Calcs. by : W.McK.
Part of Structure : Freestanding Wall	Checked by :
Calc. Sheet No. : 3 of 5	Date :

References	Calculations	Output
	The wall panel should also be checked for spanning horizontally between the piers considering failure perpendicular to the beds. The value of the bending moment can be estimated as $\dfrac{\gamma_f W_k L^2}{10}$	
Clause 22	$\gamma_f \;=\; 1.2$ Design bending moment $\;=\; \dfrac{1.2 \times W_k \times 1.8^2}{10}$ $\;=\; 0.389\, W_k \text{ kNm}$	
Table 3	$f_{kx\,perp} \;=\; 0.9 \text{ N/mm}^2$ Design moment of resistance $\;=\; \dfrac{f_{kx\,perp}}{\gamma_m} Z$	
Table 4	$\gamma_m \;=\; 3.1$ $Z \;=\; \dfrac{1000 \times 102.5^2}{6} \;=\; 1.751 \times 10^6 \text{ mm}^3$ Design moment of resistance $\;=\; \dfrac{0.9 \times 1.751 \times 10^6}{3.1 \times 10^6}$ $\;=\; 0.508 \text{ kNm}$ $0.389\, W_k \;\le\; 0.508$ $\therefore\; W_k \;\le\; 1.31 \text{ kN/m}^2 >$ pier strength	$W_k \le 0.51 \text{ kN/m}^2$
	Case (c) –Wall with Piers	
Clause 36.4.3	The outstanding flange length from the face of a pier is equal to: $(6 \times$ thickness of the wall forming the flange$)\;=\;(6 \times 102.5)$ $\;=\; 615 \text{ mm}$	

References	Calculations	Output
	$I_{xx} = \left[\left(\dfrac{328\times665^3}{12}\right) + 2\times\left(\dfrac{615\times102.5^3}{12}\right)\right]$	
	$I_{xx} = 8149 \times 10^6 \text{ mm}^4$	
	Average I_{xx} $= \dfrac{8149\times10^6}{1.8}$ $= 4530\times10^6 \text{ mm}^4$	
	Average Z_{xx} $= \dfrac{4530\times10^6}{332.5}$ $= 13.68\times10^6 \text{ mm}^3$	
	Design moment of resistance $= (0.146\times13.68\times10^6)/10^6$ $= 1.99 \text{ kNm/metre run}$	
	Design moment $= 3.75\,W_k \leq 1.99 \therefore W_k \leq 0.53 \text{ kN/m}^2$	
	The wall spanning between the piers is the same as before: $\qquad W_k \leq 1.96 \text{ kN/m}^2 > \text{pier strength}$	$W_k \leq 0.53 \text{ kN/m}^2$
	Case (d) – Chevron Wall	
see Appendix A in this text	$Z_{xx} = \dfrac{bd\left(b^2\sin^2\alpha + d^2\cos^2\alpha\right)}{6\left(b\sin\alpha + \cos\alpha\right)}$; **Note:** $\alpha = 60^0$	
	$Z_{xx} = \dfrac{102.5\times2078.5\times\left[\left(102.5^2\times0.866^2\right)+\left(2078.5^2\times0.5^2\right)\right]}{6\times\left[\left(102.5\times0.866\right)+\left(2078.5\times0.5\right)\right]}$	
	Average Z_{xx} $= \dfrac{34\times10^6}{1.8}$ $= 19.03\times10^6 \text{ mm}^3$	
	Design moment of resistance $= (0.138\times19.03\times10^6)/10^6$ $= 2.63 \text{ kNm/metre run}$	
	Design moment $= 3.75\,W_k \leq 2.63 \quad \therefore W_k \leq 0.7 \text{ kN/m}^2$	$W_k \leq 0.7 \text{ kN/m}^2$

Contract : **Leisure Park**   **Job Ref. No. : Example 3.4**
Part of Structure : **Freestanding Wall**
Calc. Sheet No. : **4 of 5**

Calcs. by : **W.McK.**
Checked by :
Date :

References	Calculations	Output

<table>
<tr>
<td>

Contract : Leisure Park　　　**Job Ref. No. :** Example 3.4
Part of Structure :　Freestanding Wall
Calc. Sheet No. : 5 of **5**

</td>
<td>

Calcs. by : W.McK.
Checked by :
Date :

</td>
</tr>
</table>

Note:
In each case consider the ratio of the maximum value of W_k and the cross-sectional area of brickwork required for 1.8 m length of wall:

Case (a)　Solid Wall
Cross-sectional area　$=$　$0.589 \, m^2$　　　$W_k =$　$0.7 \, kN/m^2$

$$\frac{W_k}{Area} = \frac{0.7}{0.589} = 1.19 \, kN/m^4$$

Case (b)　Staggered Wall
Cross-sectional area　$=$　$0.257 \, m^2$　　　$W_k =$　$0.51 \, kN/m^2$

$$\frac{W_k}{Area} = \frac{0.51}{0.257} = 1.98 \, kN/m^4$$

Case (c)　Wall with Piers
Cross-sectional area　$=$　$0.369 \, m^2$　　　$W_k =$　$0.53 \, kN/m^2$

$$\frac{W_k}{Area} = \frac{0.53}{0.369} = 1.44 \, kN/m^4$$

Case (a)　Chevron Wall
Cross-sectional area　$=$　$0.213 \, m^2$　　　$W_k =$　$0.7 \, kN/m^2$

$$\frac{W_k}{Area} = \frac{0.7}{0.213} = 3.28 \, kN/m^4$$

The efficiency of each configuration relative to the solid wall in terms of material used is:

　1.0

　1.66

　1.21

　3.28

In Example 3.4 the strength of each wall was used to determine the maximum wind load which could be resisted. The flexural compressive strength and the shear strength at the base of the wall should also be checked; in most cases these will not be critical. If desired, the compressive strength can be checked as in previous examples and the shear strength can be checked in accordance with Clause 25.

3.7   Shear Strength   (Clause 25)

In Clause 25 the characteristic shear strength of masonry, f_v , in the horizontal direction of the horizontal plane is given as:

$$f_v = (0.35 + 0.6g_A) \text{ N/mm}^2 \quad \text{For mortar designation}$$
$$\leq 1.75 \text{ N/mm}^2 \qquad \quad \text{Types (i) and (ii)}$$

$$f_v = (0.15 + 0.6g_A) \text{ N/mm}^2 \quad \text{For mortar designation}$$
$$\leq 1.4 \text{ N/mm}^2 \qquad \quad \text{Types (iii) and (iv)}$$

where g_A is the design vertical load per unit area of wall cross-section due to the vertical loads calculated from the appropriate loading conditions specified in Clause 22.

The characteristic shear strength, f_v, of bonded masonry in the vertical direction of the vertical plane may be taken as:

$$f_v = 0.7 \text{ N/mm}^2 \qquad \begin{array}{l}\text{For mortar designation} \\ \text{Types (i) and (ii)}\end{array}$$

$$f_v = 0.5 \text{ N/mm}^2 \qquad \begin{array}{l}\text{For mortar designation} \\ \text{Types (iii) and (iv)}\end{array}$$

$$f_v = 0.35 \text{ N/mm}^2 \qquad \begin{array}{l}\text{For dense aggregate} \\ \text{solid concrete blocks} \\ \text{with a minimum} \\ \text{strength of 7 N/mm}^2 \\ \text{with mortar designation} \\ \text{Types (i), (ii) and (iii)}\end{array}$$

Consider case (a) in Example 3.4:

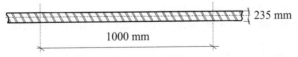

235 mm

1000 mm

Height of wall = 2.5 m, γ_m = 3.1, γ_f = 1.2   (for wind loads)
Brickwork is built in mortar designation Type (ii)
Design vertical load/unit area g_A = 0.049 N/mm^2 (assuming γ_f = 0.9 for self-weight)

Characteristic shear strength = f_v = $[0.35 + 0.6g_A]$ and $\leq$ 1.75 N/mm²
$= [0.35 + (0.6 \times 0.049)]$
$= 0.38$ N/mm²

Maximum design shear force $= \dfrac{f_v \times Area}{\gamma_m} = \left(\dfrac{0.38 \times 327.5 \times 1000}{3.1 \times 10^3}\right) = 40.1$ kN

Characteristic wind load $= W_k = \dfrac{Design\ shear\ force}{Height \times 1.0 \times \gamma_f} = \dfrac{40.1}{2.5 \times 1.2} = 13.4$ kN/m²

This value is clearly very much higher than the value, 0.7 kN/m², obtained when considering flexure.

In cases where the strength of the masonry is sufficient to transmit the required flexural tension (e.g. where a flexible damp-proof course is present) a freestanding wall must be designed as a gravity structure considering stability. In this case the overturning moment due to the applied lateral load is resisted by the self-weight stabilising moment. The design moment of resistance per unit length is based on an assumed cracked section with a narrow stress block as shown in Figure 3.16 and can be determined as shown.

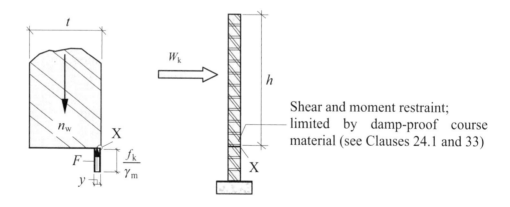

Figure 3.16

where:
y is the assumed width of the rectangular stress block,
X is the assumed point of rotation of the wall,
n_w is the design vertical load per unit length of wall (normally assumed to be 0.9 × the dead load),
F is the force per unit length due to the rectangular stress block,
h is the height of the wall,
t, W_k, f_k and γ_m are as before.

Consider rotational equilibrium about point X:

$$\text{stabilising moment} \quad \geq \quad \text{overturning moment}$$

$$\left[\left(n_w \times \frac{t}{2}\right) - \left(F \times \frac{y}{2}\right)\right] \quad \geq \quad \left[W_k \times \gamma_f \times \frac{h}{2}\right]$$

$$\text{Force } F = (\text{stress} \times \text{area}) = \left(\frac{f_k}{\gamma_m} \times y \times 1.0\right) = \frac{f_k y}{\gamma_m}$$

$$\left[\left(\frac{n_w t}{2}\right) - \left(\frac{f_k y^2}{2\gamma_m}\right)\right] \quad \geq \quad \left[\frac{W_k \gamma_f h}{2}\right] \tag{1}$$

Consider vertical equilibrium:

$$n_w = F = \frac{f_k y}{\gamma_m}$$

$$y = \frac{n_w \gamma_m}{f_k} \tag{2}$$

Substituting for equation (2) in equation (1) gives:

$$\left[\left(\frac{n_w t}{2}\right) - \left(\frac{f_k n_w^2 \gamma_m^2}{2\gamma_m f_k^2}\right)\right] \quad \geq \quad \left[\frac{W_k \gamma_f h}{2}\right]$$

$$\frac{n_w}{2}\left(t - \frac{n_w \gamma_m}{f_k}\right) \quad \geq \quad \left[\frac{W_k \gamma_f h}{2}\right]$$

The expression $\dfrac{n_w}{2}\left(t - \dfrac{n_w \gamma_m}{f_k}\right)$ represents the design moment of resistance and is given in Clause 36.5.3 of the code.

3.8   Example 3.5   Balustrade Wall

A balustrade wall in a spectators gallery above a squash court is required to resist a lateral line load applied at a height of 1.1 m above the floor as indicated in Figure 3.17. Using the design data given check the suitability of the concrete blocks and mortar combination indicated.

Design data:
Characteristic line load applied 1.1 m above the level of the floor   0.74 kN/m length
Category of manufacturing control   special
Category of construction control   special
Characteristic unit weight of the concrete blocks   20.0 kN/m^3
The wall is constructed from 390 mm thick solid concrete blocks having a ratio of height to least horizontal dimension equal to 0.6, a compressive strength of 40 N/mm^2 and set

with mortar designation Type (ii).

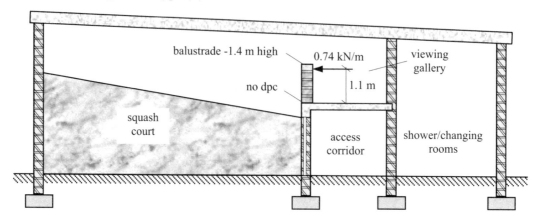

balustrade -1.4 m high
0.74 kN/m
viewing gallery
no dpc
1.1 m
squash court
access corridor
shower/changing rooms

Figure 3.17

3.8.1 Solution to Example 3.5

Contract : Squash Court **Job Ref. No. :** Example 3.5	**Calcs. by :** W.McK.
Part of Structure : BalustradeWall	**Checked by :**
Calc. Sheet No. : 1 of **5**	**Date :**

References	Calculations	Output
BS 5628 : Part 1	Structural use of unreinforced masonry	
	Consider a 1.0 m length of wall:	
Clause 22	Characteristic lateral load = 0.74 kN/m γ_f = 1.6 Design lateral load = (1.6×0.74) = 1.184 kN/m Design bending moment = (1.184×1.1) = 1.30 kNm	
Clause 22	Characteristic dead load = $(20 \times 0.39 \times 1.4)$ = 10.92 kN/m γ_f = 0.9 Design dead load = n_w = (10.92×0.9) = 9.83 kN/m	
Clause 36.5.3	Design moment of resistance = $\dfrac{n_w}{2}\left(t - \dfrac{n_w\gamma_m}{f_k}\right)$	
Table 4 Clause 23.1.6 Table 2(b)	γ_m = 2.5 f_k = 9.4 N/mm^2 $(= 9.4 \times 10^3$ kN/m$^2)$	**Adopt 390 mm thick concrete blocks with compressive strength 40 N/mm^2 in mortar designation Type (ii)**
	Design moment of resistance = $\dfrac{9.83}{2}\left(0.39 - \dfrac{9.83\times2.5}{9.4\times10^3}\right)$ = 1.9 kNm > 1.30 kNm	

3.9   Walls Containing Openings

The design guidance given in Clause 36 of the code is related to laterally loaded panels in which there are no openings. In walls where substantial openings occur or which are of irregular shape the recommendations given in Appendix D of the code can be used; they are:

*'...it will often be possible to divide them into sub-panels, which can then be calculated using the rules given in Clause **36** (see figure 26 – [Figure 3.18 in this text]). Alternatively, an analysis, using a recognised method of obtaining bending moments in flat plates, e.g. finite element or yield line, may be used and these can then be used instead of the moments obtained from the coefficients given in table 9.*

Small openings in panels will have little effect on the strength of the panel in which they occur, and they can be ignored. When suitable timber or metal frames are built into openings, the strength of the frame, taken in conjunction with the masonry panel, will often be sufficient to replace the strength lost by the area of the opening. Such cases will have to be decided by the designer, as guidance is beyond the scope of this code.'

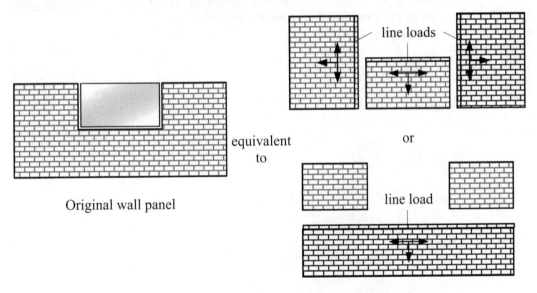

Dividing a panel into parts to allow for openings

Figure 3.18

A more rigorous approach can be used employing plate analysis techniques such as Johansen's Yield Line theory (48), to determine the design bending moments. This method was not developed for brittle materials such as masonry, it does, however, give good correlation with experimental data when applied to panels without openings. The design of panels with openings is outwith the scope of this text and further information can be found in reference (30).

3.10   Review Problems

3.1   Define the 'orthogonal ratio' in terms of characteristic flexural strength.
       (see section 3.1)

3.2   Explain the effect of pre-compression on the flexural strength of masonry.
       (see section 3.1)

3.3   Identify the factors which influence the bending moment coefficient 'α'
       for laterally loaded panels.
       (see section 3.1.1)

3.4   Explain the conditions under which the edges of a panel in a cavity wall
       can be assumed to be continuous.
       (see section 3.1.2)

3.5   Indicate how the cross-sectional geometry of a single-leaf free-standing
       wall can be modified to make it more efficient.
       (see section 3.5)

4. Reinforced and Prestressed Masonry

Objective: *'To illustrate the requirements for the limit-state design of reinforced and prestressed masonry.'*

4.1   Introduction

One of the main disadvantages of masonry as a structural material is its very low tensile strength. This deficiency frequently results in an inability to fully utilise the high compressive strength available. In a similar manner to concrete this can be overcome by adopting either:

- ◆ *Reinforced* masonry in which steel bars are introduced to resist the tensile stresses,   or
- ◆ *Prestressed* masonry in which forces are introduced to eliminate the tensile stresses.

The design requirements for both techniques are given in BS 5628 : **Part 2** : 2000. The extent of the research data available on the behaviour of reinforced and prestressed elements is still relatively limited and consequently its use to date has been restricted. The methods of design as recommended in the code are similar to those adopted for concrete. In Clause 7.1.1.2, designers are encouraged to consider '......*whether the proportion of concrete infill in a given cross-section is such that the recommendations of BS 8110 : Part 1 would be more appropriate...*'. No further advice is given but it seems reasonable to consider an element in which less than 50% of the cross-section is masonry to be a reinforced concrete element and design it accordingly.

An introduction to the design of reinforced and prestressed concrete design is given in this chapter and the reader is referred to publications (28, 30 and 32) for further, more comprehensive and detailed design advice.

4.2   Reinforced Masonry

In reinforced masonry advantage can be taken of the voids created when using particular bonds and/or the bed joints as shown in Figure 1.4 of Chapter 1. The use of bed reinforcement enhances the capacity of walls to resist lateral loading.

The limit state philosophy as used with unreinforced masonry is used for design with the appropriate partial safety factors (γ_f) for ultimate and service loads given in Clauses 7.5.2.1 and 7.5.3.1 respectively.

The partial safety factors for material strengths are given in Clause 3.5.2.2 as:

γ_{mm} for the compressive strength of masonry: (Table 7: equal to 2.0 for the *special* category of manufacturing control and 2.3 for the *normal* category of manufacturing control),

γ_{mv} for the shear strength of masonry: (Table 8: equal to 2.0),

γ_{ms} for the strength of the steel: (Table 8: equal to 1.15),

γ_{mb} for the bond strength between the infill concrete or mortar and steel:

(Table 8: equal to 1.5).

In unreinforced masonry the partial safety factor (γ_{mm}) is based on two possible categories of both manufacturing and construction control. In reinforced/prestressed masonry it is assumed that **ALL** construction will be carried out under special control and consequently Table 7 only refers to the category of manufacturing control. As indicated in Clause 7.5.3.2, when considering deflections, stress calculations or crack widths at serviceability, γ_{mm} for masonry should be taken as 1.5 and γ_{ms} for steel as 1.0

Normally only mortar designation Types (i) and (ii) as given in Table 1, are used with the exception of walls incorporating bed joint reinforcement where Type (iii) may be used. The concrete infill is specified in Clause 6.9 in terms of proportions by volume of materials, prescribed mixes or designed mixes. In all cases the maximum size of aggregate should not exceed (cover to reinforcement minus 5.0 mm). In addition, where the mix is specified by volumes the maximum size should not exceed 10 mm.

The characteristic strength of masonry 'f_k' used in the design of flexural members is given in Figures 1(a) to (d) and in Tables 3(a) to (d). The partial safety factor γ_{mm} includes an allowance for compression due to flexure, which is different from compression due to pure axial loading. The value of f_k adopted is influenced by the type, aspect ratio and compressive strength of the unit, the mortar designation and the direction of the compressive force relative to the bed face of the unit. Where compression is perpendicular to the unit bed face the value is determined as indicated in Clauses 7.4.1.1.3(a) to (g) and where the force is parallel to the bed face as indicated in Clauses 7.4.1.1.4(a) to (c).

The durability and hence cover requirements for the reinforcing steel are given in Section 10 of the code. Exposure conditions E1 to E4 are defined in Clause 10.1.2.2 and minimum cover requirements for carbon steel reinforcement are given in Table 15 as shown in Figure 4.1.

Exposure Classification	Minimum Concrete Cover To Carbon Steel Reinforcement (mm)				
	Concrete Grade (N/mm²)				
	C30	C35	C40	C45	C50
E1[b]	20	20	20[c]	20[c]	20[c]
E2	–	35	30	25	20
E3	–	–	40	30	25
E4	–	–	–	60[d]	50
Note: see Code for notes re – b, c, d, minimum cement content and maximum free water cement ratio.					

Figure 4.1

The design of flexural elements requires the evaluation of the design moment of resistance and the shear resistance at the ultimate limit state and checks on the deflection and cracking at the serviceability limit state.

4.2.1   Design Moment of Resistance   (Clause 8.2.4)

The design to resist bending is based on the principles of strain compatibility; i.e. consideration of the strain profile throughout the depth of a section, the consequent stress distribution obtained from the stress–strain diagrams and the resulting evaluation of the compressive and tensile forces and moment of resistance.

The cross-section of a beam subject to pure bending and the resulting strain diagram with typical stress–strain curves for masonry and steel are shown in Figure 4.2.

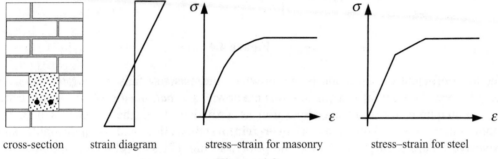

cross-section   strain diagram   stress–strain for masonry   stress–strain for steel

Figure 4.2

During bending, the top fibres are in compression (assuming a simply supported span) and the bottom fibres are in tension leading to a triangular strain diagram. To produce the profile of the stress distribution for the masonry in compression requires determination of the values of stress corresponding to the values of strain from the stress-strain diagram as shown in Figure 4.3.

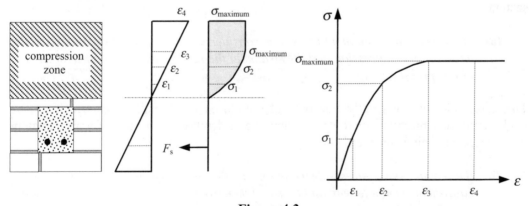

Figure 4.3

The rectangular/parabolic stress block produced can be used to evaluate the compressive force in the masonry.

In a similar manner the force in the steel can be determined by obtaining the steel stress from the stress–strain diagram and calculating the product (stress × area). The moment of resistance can be determined on the basis of the steel force or the concrete force as shown in Figure 4.4, where the tensile strength of the masonry is ignored.

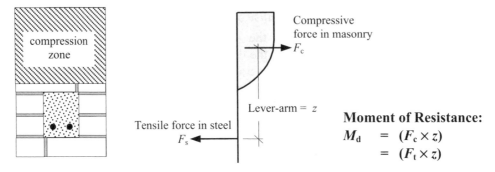

Figure 4.4

The most efficient cross-section is one in which the masonry fails in compression at the same time as the steel fails in tension, this is known as a '*balanced section*'. If **more** steel than is required for a balanced section is provided, then the section will fail in compression; this is known as an **over-reinforced section** and is undesirable since explosive failure occurs with little or no prior warning. If **less** steel than is required for a balanced section is provided, then the section will fail by yielding of the steel in tension; this is known as an **under-reinforced** section. This type of failure is desirable since warning signs such as excessive deflection or opening cracks will develop indicating that the steel is overstressed.

The design equations in BS 5628 : Part 2 : 2000 are formulated to produce under-reinforced sections. The following assumptions are made in Clause 8.2.4.1 of the code when applying strain compatibility to determine the design moment of resistance of a section:

   (a) *plane sections remain plane when considering the strain distribution in the masonry in compression and the strains in the reinforcement, whether in tension or compression;*

This assumption is not strictly true but only produces very small errors which can be neglected except in cases where deep beams, i.e. a depth-to-span ratio of approximately 0.5 or more, are being considered.

   (b) *the compressive stress distribution in the masonry is represented by an equivalent rectangle with an intensity taken over the whole compression zone of:*
$$f_k / \gamma_{mm}$$
   where:
   (c) *f_k is obtained from 7.4.1.2   and*
   γ_{mm} is given the value appropriate to the limit state being considered (see 7.5);

This assumption simplifies the analysis by replacing the rectangular/parabolic stress block shown in Figure 4.3 by an equivalent rectangular block. The characteristic compressive strength of brickwork in bending (as opposed to pure compression) is reflected in the values adopted for γ_{mm} i.e. 2.0 and 2.3 compared to 2.5 and 2.8 in Part 1 of the Code.

  (d) *the maximum strain in the outermost compression fibre at failure is 0.0035;*

  (e) *the tensile strength of the masonry is ignored;*

  (f) *the characteristic strength of the reinforcing steel is taken from Table 4, and the stress-strain relationship is taken from Figure 2;* (of the code)

The stress–strain relationship for the steel is shown in Figure 4.5 in which the values of strain at which yield is assumed to occur for mild steel (250 N/mm^2) and high yield steel (460 N/mm^2) are 0.0033 and 0.0043 respectively as indicated in BS 4449 : 1997 'Specification for hot rolled steel bars for the reinforcement of concrete'.

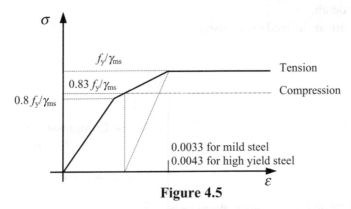

Figure 4.5

  (g) *the span to effective depth ratio of a member is not less than 1.5.*

This assumption is included since very deep beams do not behave as beams but instead as tied arches or 'wall beams' as shown in Figure 4.6. Tied arches can be designed on the basis of tension reinforcement providing the whole of the tensile force calculated assuming a lever arm equal to two-thirds of the depth, with a maximum value equal to (0.7 × span), as indicated in Clause 8.2.4.2.2 of the code and in Figure 4.6.

Area of steel $A_s = \dfrac{(M \times \gamma_{ms})}{(f_y \times lever\ arm)}$

where M is the applied bending moment

Lever-arm $= 0.67d$
$\leq 0.7L$

Span = L

Figure 4.6

The design formulae (*based on a rectangular stress block*) to determine the resistance moment for a singly reinforced rectangular member is given in Clause 8.2.4.2 as:

$$M_d = \frac{A_s f_y z}{\gamma_{ms}} \qquad \text{based on the steel strength} \qquad (1)$$

$$\leq 0.4 \frac{f_k b d^2}{\gamma_{mm}} \qquad \text{based on the concrete strength} \qquad (2)$$

where:

z is the lever arm given by $\qquad z = d \times \left(1 - \dfrac{0.5 A_s f_y \gamma_{mm}}{b d f_k \gamma_{ms}} \right)$

$$\leq 0.95d \qquad (3)$$

A_s is the cross-sectional area of the primary reinforcing steel,
b is the width of the section,
d is the effective depth,
f_k, f_y, γ_{mm} and γ_{ms} are as defined previously.

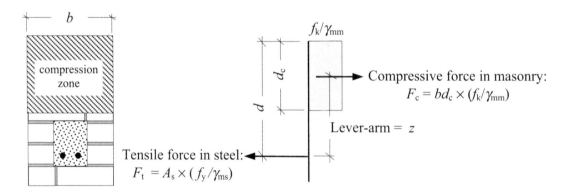

Figure 4.7

Since equations (1) and (3) both contain the variables 'A_s' and 'z' they must be solved using an iterative procedure to determine the area of steel for a given design moment. This is illustrated in Example 4.1.

   In Clause 8.2.4.2.2 an alternative procedure is given which can be used to determine the lever arm 'z' and subsequently the required area of reinforcing steel. The design moment of resistance M_d is expressed in terms of a moment resistance factor 'Q' such that:

$$M_d = Q b d^2 \qquad \text{or} \qquad Q = \frac{M}{b d^2}$$

The moment of resistance factor is defined as:

$$Q = 2c(1 - c) f_k / \gamma_{mm}$$

where:
c is the lever-arm factor $= z/d$
f_k and γ_{mm} are as defined previously
the relationship between Q, c and f_k / γ_{mm} is also given in Table 11 and Figure 3 of the code (see Figures 4.8 and 4.9).

Extract from Table 11

Table 11. Values of the moment of resistance factor, Q, for various values of f_k / γ_{mm} and lever-arm factor c				
f_k / γ_{mm}	Values of Q (N/mm²)			
c	1	5	6	20
0.95	0.095	0.475	0.570	1.900
0.85	0.255	1.275	1.530	5.100
0.84	0.269	1.344	1.613	5.376
0.83	0.282	1.411	1.693	5.644
0.82	0.295	1.476	1.771	5.904
0.72	0.403	2.016	2.419	8.064

Figure 4.8

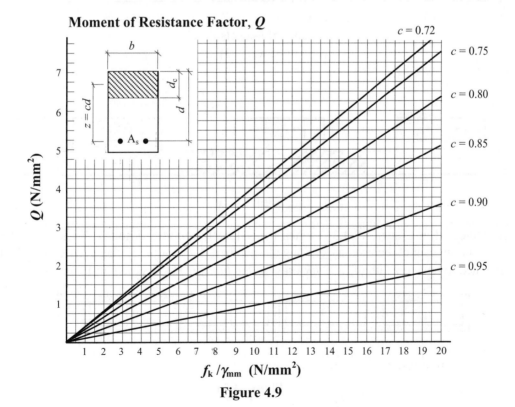

Figure 4.9

For any given value of design moment of resistance the value of Q can be calculated, and using this value with f_k / γ_{mm} and either Table 11 or Figure 3 the lever-arm factor 'c' ($= z/d$) can be determined from which 'z' can be found.

In pocket type walls where the reinforcement is concentrated as shown in Figure 4.10, the cross-section can be considered to be a '*flanged beam*' in which the flange thickness 't_f' should be taken as the thickness of the masonry but no greater than the '$(0.5 \times d)$' as indicated in Clause 8.2.4.3.1. In addition, the width of the flange should be taken as the least of:

(a)   *(the width of the pocket or rib)* + *(12 × the thickness of the flanges),*
(b)   *the spacing of the pockets or ribs,*
(c)   *one-third the height of the wall.*

Note: where the spacing of the ribs exceeds 1.0 m, the ability of the masonry to span horizontally between the ribs should also be checked.

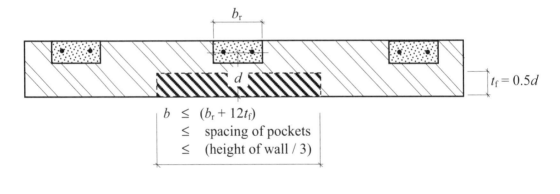

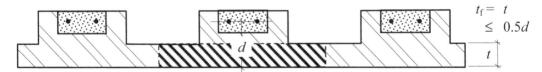

Figure 4.10

As indicated in Clause 4.2.4.3.2, when using locally reinforced hollow brickwork where the section cannot behave as a flanged member, the width of the section should be assumed to have a width of *(3 × thickness of the blockwork)* as shown in Figure 4.11.

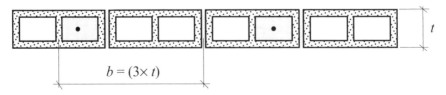

Figure 4.11

The design formulae to determine the resistance moment of flanged sections are:

$$M_d = \frac{A_s f_y z}{\gamma_{ms}} \qquad \text{based on the steel strength as before} \qquad (4)$$

$$\leq \frac{f_k b t_f}{\gamma_{mm}}(d - 0.5t_f) \quad \text{based on the concrete strength} \qquad (5)$$

$$z = d \times \left(1 - \frac{0.5 A_s f_y \gamma_{mm}}{b d f_k \gamma_{ms}}\right) \leq 0.95d \qquad (6)$$

where all variables are as defined previously.

4.2.2   Design Shear Resistance   (Clause 8.2.5)

The design shear resistance of a section can be calculated on the basis of the average design shear strength (f_v / γ_{mv}) where f_v is the characteristic shear strength as given in Clause 7.4.1.3 and γ_{mv} is the partial safety factor.

The characteristic shear strength for reinforced sections subject to bending is defined in Clause 7.4.1.3.1.a) '*For reinforced sections in which the reinforcement is placed in bed or vertical joints, including Quetta bond and other sections where the reinforcement is wholly surrounded with mortar designation (i) or (ii)*' as:

$$f_v = 0.35 \text{ N/mm}^2$$

*For simply supported beams or cantilevers where the ratio of the shear span 'a' (see **3.6**) to the effective depth is less than 2, f_v may be increased by a factor $2d/a_v$*
where:
d is the effective depth,
a_v is the distance from the face of the support to the nearest edge of a principal[1] load,
a shear span $= \dfrac{maximum\ design\ bending\ moment}{maximum\ design\ shear\ force}$

provided that f_v is not taken as greater than 0.7 N/mm^2.

Similarly characteristic shear strength is defined in Clause 7.4.1.3.1.b): '*For reinforced sections in which the main reinforcement is placed within pockets, cores or cavities filled with concrete infill as defined in Clause 6.9.1, the characteristic shear strength of the masonry, f_v*' as:

$$f_v = 0.35 + 17.5\rho$$

where:
$\rho = A_s/db$,
A_s is the cross-sectional area of primary reinforcing steel,
b and d are as defined previously.
provided that f_v is not taken as greater than 0.7 N/mm^2.

---

[1] A principal load is defined in Clause 8.2.5.3 as any concentrated load (or loads), which contributes more than 70% of the total shear force at a support.

'For simply supported beams or cantilever retaining walls where the ratio of the shear span 'a' to the effective depth 'd' is 6 or less, f_v may be increased by a factor $\{2.5 - 0.25(a/d)\}$ provided that f_v is not taken to be greater than 1.75 N/mm^2.'

In Clause 8.2.5, the average shear stress due to the design loads is given by:

$$v = \frac{V}{bd}$$

where:
v is the average design shear stress,
V is the shear force due to the design loads,
d is the effective depth (or for a flanged member, the actual thickness of the masonry between the ribs, if this is less than the effective depth as defined in **3.7**).

If $v < f_v/\gamma_{mv}$ then shear reinforcement is not required (**see *Note**)
If $v > f_v/\gamma_{mv}$ then shear reinforcement should be provided such that:

$$\frac{A_{sv}}{s_v} \geq \frac{b(v - f_v/\gamma_m)\gamma_{ms}}{f_y}$$

where:
A_{sv} is the cross-sectional area of reinforcing steel resisting shear forces,
s_v is the spacing of shear reinforcement along the member, provided that it is not taken to be greater than $0.75d$,
v is the shear stress due to design loads, provided that it is not taken to be greater than $2.0/\gamma_m$ N/mm^2,
b, f_v, f_y, γ_{ms}, γ_{mv}, are as defined previously.

*Note: In these circumstances, because of the sudden nature of shear failure designers are advised to consider the use of nominal links when designing beams. The requirements for nominal links as given in Clause 8.6.5.2 are:

$$\frac{A_{sv}}{s_v} \geq 0.002b_t \qquad \text{when using mild steel}$$

or

$$\frac{A_{sv}}{s_v} \geq 0.0012b_t \qquad \text{when using high yield steel}$$

where:
A_{sv} is the cross-sectional area of reinforcing steel resisting shear forces,
b_t is the width of the beam at the level of the tension reinforcement.

4.2.3 *Deflection and Cracking (Clauses 8.2.3, 8.2.6 and 8.2.7)*
The deflection of a reinforced beam may be estimated using the guidelines given in Appendix C, however, the accuracy of such detailed calculations are open to question. In most cases, i.e. all but the most stringent serviceability requirements beyond those

specified in Clause 7.1.2.2, the limiting ratios given in Tables 9 and 10 (see Figures 4.12 4.13) for walls and beams respectively may be used to limit deflection and cracking.

Table 9. Limiting ratios of span to effective depth for laterally-loaded walls	
End Condition	**Ratio**
Simply supported	35
Continuous or spanning in two directions	45
Cantilever with values of ρ up to and including 0.005	18

Table 10. Limiting ratios of span to effective depth for beams	
End Condition	**Ratio**
Simply supported	20
Continuous	26
Cantilever	7

Figure 4.12   **Figure 4.13**

4.2.4   *Effective Span and Lateral Restraint of Beams*

The effective span of elements and the restraint requirements to ensure that instability caused by lateral torsional buckling does not occur are defined in Clauses 8.2.2 and 8.2.3 respectively.

For simply supported or continuous beams:

effective span $\leq$ the distance between the centres of supports
$\leq$ (the clear distance between the supports + d)

clear distance between lateral restraints $\leq$ $60\, b_c$
$\leq$ $250\, b_c^2 / d$

where: b_c is the width of the compression face midway between restraints,
d is the effective depth.

For cantilevers:

effective span $\leq$ the distance between the end of the cantilever and the centre of its support;
$\leq$ the distance between the end of the cantilever and the face of the support + $d/2$

For cantilevers with lateral restraint provided only at the support:
clear distance from the end of the cantilever to the face of the support $\leq$ $25\, b_c$
$\leq$ $100\, b_c^2 / d$

Provided no damage will be caused to any applied finish due to deflection or cracking, the values given in Table 9 for walls may be increased by 30% for free-standing walls not forming part of a building and subjected predominantly to wind loads.

4.2.5 Structural Detailing

The detailing of reinforced masonry is similar to that for reinforced concrete, however, the wide variety of structural units available and possible forms of construction result in guidelines which are less prescriptive than those given in BS 8110 for concrete. BS 5628 : Part 2 : 2000 provides detailing advice and rules in Clause 8.6, the main elements of which are summarised in Table 4.1.

Detailing Requirements (Clause 8.6)		
Detail	**Clause No.**	**Main Provisions**
Area of main reinforcement	*8.6.1*	No minimum % of reinforcement is given but consideration should be given to considering a section as *unreinforced (i.e. use Part 1 of the code)* when the area of reinforcement is a small proportion of the gross area of the section.
Maximum size of reinforcement	*8.6.2*	main bar diameter $\leq$ 6 mm when placed in joints, $\leq$ 32 mm in pocket type walls, $\leq$ 25 mm elsewhere.
Minimum area of secondary reinforcement	*8.6.3*	Area $\geq 0.05\%$ bd in one-way spanning walls and slabs. Secondary steel may be omitted from pocket-type walls except where specifically required to tie masonry to the infill concrete. Some or all of the reinforcement may be used to help control cracking due to shrinkage or expansion, thermal and moisture movements.
Spacing of main and secondary reinforcement	*8.6.4*	**minimum** $s \geq$ (aggregate size + 5 mm), $\geq$ bar diameter, $\geq$ 10 mm. 's' applies to both the horizontal or vertical directions. **maximum** $s \leq$ 500 mm for tension reinforcement. The recommendations do not apply to reinforcement concentrated in cores or pockets. In vertical pockets or cores < 125 mm × 125 mm only single bars are permitted except at laps.
Links *spacing* *columns* For anchorage and curtailment see Handbook (ref: 36)	*8.2.5.1, 8.6.4 and 8.6.5.2. 8.6.5.3*	$s \leq 0.75d$ If $A_s \leq \{0.25\% \times$ Area of the masonry $(A_m)\}$, links are not required. If $A_s \geq 0.25\% A_m$ and the design axial load > 25% of the axial load resistance: diameter of links ≥ 6 mm $s \leq$ least lateral column dimension, $\leq 50 \times$ link diameter, $\leq 20 \times$ main bar diameter.

Table 4.1

4.3   Example 4.1   Reinforced Masonry Beam 1

The reinforced masonry beam indicated in Figure 4.14 is required to resist an ultimate design bending moment of 80 kNm and an ultimate design shear force of 71.1 kN. Using the design data given:

i)   check the design moment of resistance with respect to the masonry strength and determine the area of primary steel required,

ii)   check the suitability of the section with respect to shear and design any necessary reinforcement,

iii)   check the suitability of the section with respect to deflection, cracking and lateral torsional buckling.

Design data:

Category of manufacturing control for standard format brick units	normal
Unit compressive strength of bricks	50 N/mm^2
Mortar designation	Type (ii)
Infill concrete grade	35 N/mm^2
Characteristic strength of steel reinforcement	460 N/mm^2
Exposure condition	E1
Effective span (simply supported span)	4.5 m

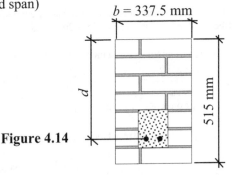

Figure 4.14

4.3.1   Solution to Example 4.1

Contract : **Job Ref. No. : Example 4.1**	Calcs. by : W.McK.
Part of Structure : Reinforced Masonry Beam 1	Checked by :
Calc. Sheet No. : **1** of **4**	Date :

References	Calculations	Output
BS 5628 : Part 2	Structural use of reinforced and prestressed masonry	
	(i) Bending Moment Resistance and area of primary steel:	
Clause 8.2.2	Resistance Moment M_d = $\dfrac{A_s f_y z}{\gamma_{ms}}$ $\leq$ $\dfrac{0.4 f_k b d^2}{\gamma_{mm}}$	

Contract :	Job Ref. No. : Example 4.1	Calcs. by : W.McK.
Part of Structure : Reinforced Masonry Beam 1		Checked by :
Calc. Sheet No. : 2 of 4		Date :

References	Calculations	Output
Table 8 Table 7 Clause 7.4.1.1.4	Partial safety factor for steel reinforcement γ_{ms} = 1.15 Partial safety factor for masonry units γ_{mm} = 2.3 f_k *where the compressive force is parallel to the bed face of the unit.* The value of f_k for masonry without holes, frogged bricks where the frogs are filled and filled hollow blocks is obtained from Table 3(a); [see Clause 7.4.1.1.3 (a)]	
Table 3(a)	Mortar designation (ii) / Unit strength 50 N/mm² f_k = 12.2 N/mm²	
Clause 10.1.2.5 Table 15	Exposure situation E1 / Concrete infill grade 35 N/mm² Minimum concrete cover to primary steel = 20 mm Assume 25 mm diameter reinforcing bars are to used. Effective depth = $(515 - 75 - 20 - 12.5)$ = 407.5 mm (i) Design moment of resistance based on masonry strength: $$M_d = \frac{0.4 f_k bd^2}{\gamma_{mm}} = \frac{\left(0.4 \times 12.2 \times 337.5 \times 407.5^2\right)}{2.3 \times 10^6}$$ = 118.9 kNm > 80 kNm (ii) Design moment of resistance based on steel strength: $$M_d = \frac{A_s f_y z}{\gamma_{ms}} \geq 80 \text{ kNm} \quad \therefore A_s \geq \frac{80 \times 10^6 \times \gamma_{ms}}{f_y z}$$ And $\quad z = d\left(1 - \frac{0.5 A_s f_y \gamma_{mm}}{bd f_k \gamma_{ms}}\right) \leq 0.95d$ Assume an initial value for z equal to 0.75d $$A_s \geq \frac{\left(80 \times 10^6 \times 1.15\right)}{460 \times (0.75 \times 407.5)} \approx 654 \text{ mm}^2$$ The value assumed for 'z' should be checked: $$z = d\left(1 - \frac{0.5 A_s f_y \gamma_{mm}}{bd f_k \gamma_{ms}}\right) \text{ and } \frac{0.5 f_y \gamma_{mm}}{bd f_k \gamma_{ms}} = 2.74 \times 10^{-4}$$ $$z = d\left[1 - (654 \times 2.74 \times 10^{-4})\right] = 0.82d$$	**Section is adequate with respect to masonry strength.**

Contract : **Job Ref. No. : Example 4.1**	**Calcs. by : W.McK.**
Part of Structure : Reinforced Masonry Beam 1	**Checked by :**
Calc. Sheet No. : 3 of 4	**Date :**

References	Calculations	Output
	Revised value of A_s $\geq \left(\dfrac{654 \times 0.75}{0.82} \right) = 598 \text{ mm}^2$ $z = d[1 - (598 \times 2.74 \times 10^{-4})] = 0.84d$ Revised value of A_s $\geq \left(\dfrac{654 \times 0.75}{0.84} \right) = 584 \text{ mm}^2$ It is evident from one more iteration that the value of steel area stabilises at 584 mm^2 and that the lever-arm 'z' is greater than the initial assumed value and less than 0.95d; the A_s value is sufficiently accurate for design. Use 2 × high yield steel 10 mm diameter bars (628 mm^2)	 **Provide 2T10 bars**
	(ii) Shear resistance and area of shear steel:	
Clause 8.2.5.1	Shear stress due to design loads $v = \dfrac{V}{bd}$ $v = \dfrac{71.1 \times 10^3}{337.5 \times 407.5} = 0.52 \text{ N/mm}^2$	
Clause 7.4.1.3.1.b	Characteristic shear strength $f_v = 0.35 + 17.5\rho$ $\leq 0.7 \text{ N/mm}^2$ $\rho = \dfrac{A_s}{bd} = \dfrac{628}{(337.5 \times 407.5)} = 0.0045$ $f_v = [0.35 + (17.5 \times 0.0045)] = 0.43 \text{ N/mm}^2$ For simply supported beams where the ratio of the shear span 'a' to the effective depth 'd', is 6 or less, f_v may be increased by a factor equal to $[2.5 - 0.25(a/d)]$ provided that f_v is not taken as greater than 1.75 N/mm^2.	
Clause 3.8	shear span $a = \dfrac{\text{maximum design bending moment}}{\text{maximum design shear force}}$ $a = \dfrac{80}{71.1} = 1.12 \text{ m}$ $a/d = \dfrac{1120}{407.5} = 2.75 \; < \; 6.0$ $f_v = \{[2.5 - 0.25(a/d)] \times 0.43\}$ $= \{[2.5 - (0.25 \times 2.75)] \times 0.43\} = 0.78 \text{ N/mm}^2$ $\leq 1.75 \text{ N/mm}^2$	

Contract : Part of Structure : Reinforced Masonry Beam 1 Calc. Sheet No. : 4 of 4 Job Ref. No. : Example 4.1	Calcs. by : W.McK. Checked by : Date :

References	Calculations	Output
Table 8	Partial safety factor for shear strength $\gamma_{mv} = 2.0$ Design shear strength $= \dfrac{f_v}{\gamma_{mv}} = \dfrac{0.78}{2.0} = 0.39 \text{ N/mm}^2$ Since the design shear strength < shear stress due to design loads shear reinforcement is required.	
Clause 8.2.5.1	$\dfrac{A_{sv}}{s_v} \geq \dfrac{b(v - f_v / \gamma_{mv})\gamma_{ms}}{f_y}$ where $s_v \leq 0.75d$ The link spacing s_v is dictated by the bonding detail, e.g. assume 2-legged links at $\approx$169 mm centres horizontal cross-section of section of beam	
Table 4 Table 8	$\gamma_{ms} = 1.15$ assume mild steel with $f_s = 250 \text{ N/mm}^2$ $A_{sv} \geq \dfrac{[337.5 \times (0.52 - 0.39)/1.15] \times 169}{250} \approx 26 \text{ mm}^2$	
Clause 8.6.5.2	Recommended nominal links $A_{sv} = 0.002 b_t s_v$ $A_{sv} = (0.002 \times 337.5 \times 169) = 114 \text{ mm}^2$ 6 mm diameter bars at 250 centres provide 113 m^2 $\therefore$ Using 6 mm bars as indicated provides $\left(113 \times \dfrac{250}{169}\right)$ $= 167 \text{ mm}^2$	Provide 6 mm diameter, mild steel links @ 169 mm spacing as indicated
Clause 8.2.3.3 Table 10	**Limiting dimensions to satisfy the serviceability limit states of deflection and cracking:** actual $\dfrac{span}{d} = \dfrac{4500}{407.5} = 11 \leq 20$ $\therefore$ adequate	Beam is adequate with respect to deflection and cracking
Clause 8.2.3.3	**To ensure lateral stability:** clear distance between lateral restraints: $\leq 60\,b_c \leq (60 \times 337.5)/10^3 = 20.25 \text{ m}$ $\leq 250\,b_c^2/d \leq (250 \times 337.5^2)/(407.5 \times 10^3) = 69.8 \text{ m}$ Actual distance between restraints $= 4.5 \text{ m} << 20.25 \text{ m}$	Beam proportions are adequate to prevent lateral torsional buckling

An alternative solution to using the equations given for bending moment of resistance is to use the design table or chart given in the code to determine the area of steel required. Consider the beam in Example 4.1

$M_d \geq 80$ kNm, $\quad b = 337.5$ mm, $\quad d = 407.5$ mm, $\quad f_k = 12.2$ N/mm^2, $\quad \gamma_{mm} = 2.3$

$$Q = \frac{M}{bd^2} = \frac{80 \times 10^6}{337.5 \times 407.5^2} = 1.43 \text{ N/mm}^2 \text{ ; } f_k / \gamma_{mm} = \frac{12.2}{2.3} = 5.3 \text{ N/mm}^2$$

From Table 11 or Figure 3 (see Figure 4.15 and Figure 4.16 of this text)

lever-arm factor $\quad c \approx 0.84$ and hence $z = 0.84d$ and $A_s = \dfrac{M_d \gamma_{ms}}{f_y z}$ as before.

Extract from Table 11

Table 11. Values of the moment of resistance factor, Q, for various values of f_k / γ_{mm} and lever-arm factor c

c \ f_k/γ_{mm}	Values of Q (N/mm^2)				
	1		5	6	20
0.85	0.255		1.275	1.530	5.100
0.84	0.269		**1.344**	**1.613**	5.376
0.83	0.282		**1.411**	**1.693**	5.644
0.82	0.295		1.476	1.771	5.904

Figure 4.15

Using **Figure 3** from the code:

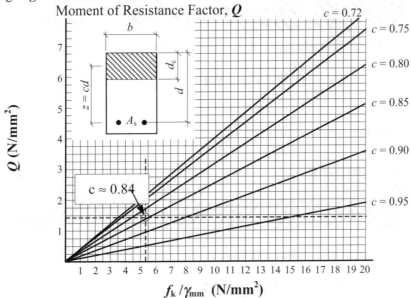

Figure 4.16

4.4   Example 4.2   Reinforced Masonry Wall 1

Three possible cross-sections for a reinforced masonry wall are indicated in Figure 4.17. Using the design data given, check the suitability of each cross-section with respect to:

    i)   the design bending moment equal to 60 kNm,
    ii) the design shear force equal to 90 kN   and,
    iii) the serviceability requirements for deflection and cracking.

Design data:

Category of manufacturing control for standard format brick units	special
Unit compressive strength of bricks	27.5 N/mm^2
Mortar designation	Type (ii)
Infill concrete grade	35 N/mm^2
Characteristic strength of steel reinforcement (f_y)	460 N/mm^2
Exposure condition	E1
Height of wall	3.5 m

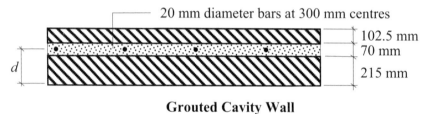

20 mm diameter bars at 300 mm centres

102.5 mm
70 mm
215 mm

d

Grouted Cavity Wall

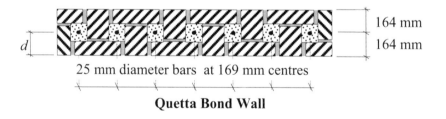

164 mm
164 mm

d

25 mm diameter bars at 169 mm centres

Quetta Bond Wall

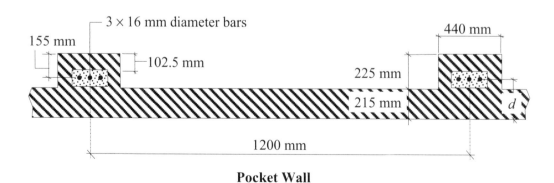

3 × 16 mm diameter bars

155 mm

102.5 mm

440 mm

225 mm

215 mm

d

1200 mm

Pocket Wall

Figure 4.17

4.4.1   Solution to Example 4.2

Contract : **Job Ref. No. :** Example 4.2	**Calcs. by :** W.McK.
Part of Structure : Reinforced Masonry Wall 1	**Checked by :**
Calc. Sheet No. : 1 of **7**	**Date :**

References	Calculations	Output
BS 5628 : Part 2	Structural use of reinforced and prestressed masonry	
	(i) Grouted Cavity Wall	
	20 mm diameter bars at 300 mm centres	
	102.5 mm 70 mm 215 mm	
Clause 8.2.2	Resistance Moment M_d $= \dfrac{A_s f_y z}{\gamma_{ms}} \le \dfrac{0.4 f_k b d^2}{\gamma_{mm}}$	
Table 8	Partial safety factor for steel reinforcement γ_{ms} = 1.15	
Table 7	Partial safety factor for masonry units γ_{mm} = 2.0	
Table 3(a)	Mortar designation (ii) / Unit strength 27.5 N/mm^2 f_k = 7.9 N/mm^2	
Clause 10.1.2.6	'.*For grouted-cavity or Quetta bond construction, the minimum cover for reinforcement for reinforcement selected using table 14 should be as follows:*...	
	(b) ***carbon steel reinforcement used in exposure situation E2:20 mm concrete;*** ...' Exposure situation E2 / Concrete infill grade 35 N/mm^2 ∴ Cover $\ge$ 20 mm	
	Actual cover = $[0.5 \times (70 - 20)]$ = 25 mm > 20 mm	**Cover is adequate**
Clause 6.9.1	Maximum aggregate size $\le$ (cover – 5 mm) = 20 mm Effective depth d = $(215 + 35 = 10)$ = 250 mm	
	$A_s = \left[\left(\dfrac{\pi \times 20^2}{4} \right) \times \left(\dfrac{1000}{300} \right) \right] =$ 1047 mm^2/m width	
Clause 8.2.4.2.1	lever-arm $z = \ d\left(1 - \dfrac{0.5 \times 1047 \times 460 \times 2.0}{1000 \times 250 \times 7.9 \times 1.15} \right)$ = $0.78d$ < $0.95d$	
	Strength based on steel $= \dfrac{A_s f_y z}{\gamma_{ms}}$	
	$= \dfrac{1047 \times 460 \times 0.78 \times 250}{1.15 \times 10^6}$ = 81.67 kNm	

References	Calculations	Output
	Strength based on concrete $= \dfrac{0.4 f_k b d^2}{\gamma_{mm}}$ $= \dfrac{0.4 \times 7.9 \times 1000 \times 250^2}{2.0 \times 10^6} = 98.75 \text{ kNm}$ **Strength is based on steel = 81.67 kNm > 60 kNm** **Shear resistance:**	**Grouted cavity wall is adequate with respect to bending**
Clause 8.2.5.1	Shear stress due to design loads $v = \dfrac{V}{bd}$ $v = \dfrac{90 \times 10^3}{1000 \times 250} = 0.36 \text{ N/mm}^2$	
Clause 7.4.1.3.1.b	Characteristic shear strength $f_v = 0.35 + 17.5\rho$ $\le 0.7 \text{ N/mm}^2$ $\rho = \dfrac{A_s}{bd} = \dfrac{1047}{(1000 \times 250)} = 0.0042$ $f_v = [0.35 + (17.5 \times 0.0042)] = 0.42 \text{ N/mm}^2$ For cantilever retaining walls where the ratio of the shear span 'a' to the effective depth 'd', is 6 or less, f_v may be increased by a factor equal to $[2.5 - 0.25(a/d)]$ provided that f_v is not taken as greater than 1.75 N/mm^2.	
Clause 3.8	shear span $a = \dfrac{maximum\ design\ bending\ moment}{maximum\ design\ shear\ force}$ $a = \dfrac{60}{90} = 0.67 \text{ m}$ $a/d = \dfrac{670}{250} = 2.68 < 6.0$ $f_v = \{[2.5 - 0.25(a/d)] \times 0.42\}$ $= \{[2.5 - (0.25 \times 2.68)] \times 0.42\} = 0.76 \text{ N/mm}^2$ $\le 1.75 \text{ N/mm}^2$	
Table 8	Partial safety factor for shear strength $\gamma_{mv} = 2.0$ Design shear strength $= \dfrac{f_v}{\gamma_{mv}} = \dfrac{0.76}{2.0} = 0.38 \text{ N/mm}^2$ $\ge 0.36 \text{ N/mm}^2$	**Grouted cavity wall is adequate with respect to shear**

Contract :

Job Ref. No. : Example 4.2

Part of Structure : Reinforced Masonry Wall 1

Calc. Sheet No. : **2** of **7**

Calcs. by : W.McK.

Checked by :

Date :

Contract :	Job Ref. No. : Example 4.2	Calcs. by : W.McK.
Part of Structure : Reinforced Masonry Wall 1		Checked by :
Calc. Sheet No. : 3 of 7		Date :

References	Calculations	Output
	Note: In the case of walls in which the shear stress due to the design loads is greater than the design shear strength it is necessary to increase the thickness of the wall. The provision of shear reinforcement in walls is generally not very practical.	
Clause 8.2.3.2	**Limiting dimensions to satisfy the serviceability limit states of deflection and cracking:**	
Table 10	Since $\rho = 0.0042 < 0.005$ $\quad \dfrac{span}{d} \leq 18$ actual $\dfrac{span}{d} = \dfrac{3500}{250} = 14 \leq 18 \therefore$ adequate	**Wall is adequate with respect to deflection and cracking**
Clause 8.6.3	Minimum area of secondary reinforcement $\geq 0.05\% bd$ $= (0.05 \times 1000 \times 250)/100 = 125 \text{ mm}^2/\text{m}$ Provide 10 mm diameter high yield steel bars at 500 mm centres ($A_s = 157 \text{ mm}^2/\text{m}$).	
Clause 8.6.4	Minimum spacing $\geq$ (aggregate size + 5 mm) $= 25 \text{ mm}$ $\geq$ bar diameter: $\quad = $ (20 mm for main steel) $\quad = $ (10 mm for secondary steel) $\geq 10 \text{ mm}$ Maximum spacing $\leq 500 \text{ mm} \quad \therefore$ spacing acceptable	**Provide T10 bars @ 500 mm centres**
	(ii) Quetta Bond Wall d 164 mm 164 mm 20 mm diameter bars at 169 mm	
Clause 8.2.2	Resistance moment $M_d = \dfrac{A_s f_y z}{\gamma_{ms}} \leq \dfrac{0.4 f_k b d^2}{\gamma_{mm}}$	
Table 8	Partial safety factor for steel reinforcement $\gamma_{ms} = 1.15$	
Table 7	Partial safety factor for masonry units $\gamma_{mm} = 2.0$	
Table 3(a)	Mortar designation (ii) / Unit strength 27.5 N/mm^2 $f_k = 7.9 \text{ N/mm}^2$	

Contract : **Job Ref. No. :** Example 4.2	**Calcs. by :** W.McK.
Part of Structure : Reinforced Masonry Wall 1	**Checked by :**
Calc. Sheet No. : 4 of 7	**Date :**

References	Calculations	Output
Clause 10.1.2.6 Table 15	As before: Exposure situation E1 / Concrete infill grade 35 N/mm^2 ∴ Cover ≥ 20 mm Actual cover = $(164 - 112.5 - 10)$ = 41 mm > 20 mm Effective depth d = 164 mm $A_s = \left[\left(\dfrac{\pi \times 20^2}{4}\right) \times \left(\dfrac{1000}{169}\right)\right]$ = 1859 mm^2/m width	**Cover is adequate**
Clause 8.2.4.2.1	lever-arm $z = d\left(1 - \dfrac{0.5 \times 1859 \times 460 \times 2.0}{1000 \times 164 \times 7.9 \times 1.15}\right)$ = $0.43d$ < $0.95d$ Strength based on steel = $\dfrac{A_s f_y z}{\gamma_{ms}}$ = $\dfrac{1859 \times 460 \times 0.43 \times 164}{1.15 \times 10^6}$ = 52.4 kNm Strength based on concrete = $\dfrac{0.4 f_k b d^2}{\gamma_{mm}}$ = $\dfrac{0.4 \times 7.9 \times 1000 \times 164^2}{2.0 \times 10^6}$ = 42.5 kNm **Strength is based on masonry = 42.5 kNm << 60 kNm**	**Quetta Bond wall is inadaequate with respect to bending**
Clause 8.2.5.1	**Shear resistance:** Shear stress due to design loads $v = \dfrac{V}{bd}$ $v = \dfrac{90 \times 10^3}{1000 \times 164}$ = 0.55 N/mm^2	
Clause 7.4.1.3.1.b	Characteristic shear strength f_v = $0.35 + 17.5\rho$ ≤ 0.7 N/mm^2 $\rho = \dfrac{A_s}{bd} = \dfrac{1859}{(1000 \times 164)}$ = 0.0113 $f_v = [0.35 + (17.5 \times 0.0113)]$ = 0.55 N/mm^2 For cantilever retaining walls where the ratio of the shear span	

Contract :	Job Ref. No. : Example 4.2	Calcs. by : W.McK.
Part of Structure : Reinforced Masonry Wall 1		Checked by :
Calc. Sheet No. : 5 of 7		Date :

References	Calculations	Output
	'a' to the effective depth 'd', is 6 or less, f_v may be increased by a factor equal to [2.5 – 0.25(a/d)] provided that f_v is not taken as greater than 1.75 N/mm².	
Clause 3.8	shear span a = $\dfrac{maximum\ design\ bending\ moment}{maximum\ design\ shear\ force}$ a = $\dfrac{60}{90}$ = 0.67 m as before a/d = $\dfrac{670}{164}$ = 4.1 < 6.0 f_v = {[2.5 – 0.25(a/d)] × 0.55} = {[2.5 – (0.25 × 4.1)] × 0.55} = 0.81 N/mm² ≤ 1.75 N/mm²	
Table 8	Partial safety factor for shear strength γ_{mv} = 2.0 Design shear strength = $\dfrac{f_v}{\gamma_{mv}}$ = $\dfrac{0.81}{2.0}$ = 0.41 N/mm² ≪ 0.55 N/mm²	**Quetta Bond wall is inadequate with respect to shear**
	Note: It is clearly evident that this thickness of wall is not suitable to resist the required design loads and the dimensions must be increased accordingly. **(iii) Pocket Wall**	

Pocket Wall

Clause 8.2.4.3	Resistance moment M_d = $\dfrac{A_s f_y z}{\gamma_{ms}}$ ≤ $\dfrac{f_k b t_f (d - 0.5 t_f)}{\gamma_{mm}}$	
Table 8	Partial safety factor for steel reinforcement γ_{ms} = 1.15	
Table 7	Partial safety factor for masonry units γ_{mm} = 2.0	
Table 3(a)	Mortar designation (ii) / Unit strength 27.5 N/mm² f_k = 7.9 N/mm²	

Contract :	Job Ref. No. : Example 4.2	Calcs. by : W.McK.
Part of Structure : Reinforced Masonry Wall 1		**Checked by :**
Calc. Sheet No. : 6 of **7**		**Date :**

References	Calculations	Output
Clause 8.2.4.3.1	Flange thickness t_f = masonry thickness (t)	
	$\leq$ 0.5d	
Figure 6	Minimum cover in pocket walls:	
	Recommended minimum cover given in Table 15	
Clause 10.1.2.6	As before:	
Table 15	Exposure situation E2 / Concrete infill grade 35 N/mm^2	
	$\therefore$ Cover $\geq$ 20 mm	
	Actual cover = $(155 - 102.5 - 8)$ = 44.5 mm > 20 mm	**Cover is adequate**
	Effective depth d = $(215 + 225 - 155)$ 285 mm	
	Flange thickness t_f = (masonry thickness) = 215 mm	
	$\leq$ 0.5d = (0.5×285) = 142.5 mm	
	$\therefore$ t_f = 142.5 mm	
Clause 8.2.4.3.1	Width of flange b $\leq$ $(b_r + 12\ t_f)$	
	$\leq$ spacing of pockets	
	$\leq$ (height of wall)/3	
	b $\leq$ $(b_r + 12\ t_f)$ = $[440 + (12 \times 142.5)]$ = 2150 mm	
	$\leq$ 1200 mm	
	$\leq$ (3500/3) = 1166.7 mm	
	$\therefore$ Flange width b = 1167.7 mm	
	A_s = $\left[\left(\dfrac{\pi \times 16^2}{4}\right)\times 3\right]$ = 603 mm^2	
Clause 8.2.4.2.1	lever-arm z = $d\left(1 - \dfrac{0.5 \times 603 \times 460 \times 2.0}{1166.7 \times 285 \times 7.9 \times 1.15}\right)$ = 0.91d	
	$<$ 0.95d	
	Strength based on steel = $\dfrac{A_s f_y\ z}{\gamma_{ms}}$	
	= $\dfrac{603 \times 460 \times 0.91 \times 285}{1.15 \times 10^6}$ = 62.5 kNm	
Clause 8.2.4.3.1	Strength based on concrete = $\dfrac{f_k b t_f (d - 0.5 t_f)}{\gamma_{mm}}$	

Contract :	Job Ref. No. : Example 4.2	Calcs. by : W.McK.
Part of Structure : Reinforced Masonry Wall 1		Checked by :
Calc. Sheet No. : 7 of 7		Date :

References	Calculations	Output
	$= \dfrac{(7.9 \times 1166.7 \times 142.5) \times [285 - (0.5 \times 142.5)]}{2.0 \times 10^6} = 140.0$ kNm	**Pocket wall is adequate with respect to bending**
	Strength is based on steel = 62.5 kNm > 60 kNm	
	Shear resistance:	
Clause 8.2.5.1	Shear stress due to design loads $\quad v = \dfrac{V}{bd}$	
	$v = \dfrac{90 \times 10^3}{1166.7 \times 285} = 0.27$ N/mm^2	
Clause 7.4.1.3.1.b	Characteristic shear strength $\quad f_v = 0.35 + 17.5\rho$ $\qquad\qquad\qquad\qquad\qquad\quad \le \ 0.7$ N/mm^2	
	$\rho = \dfrac{A_s}{bd} = \dfrac{603}{(1166.7 \times 285)} = 0.0018$	
	$f_v = [0.35 + (17.5 \times 0.0018)] = 0.38$ N/mm^2	
	For cantilever retaining walls where the ratio of the shear span 'a' to the effective depth 'd', is 6 or less, f_v may be increased by a factor equal to $[2.5 - 0.25(a/d)]$ provided that f_v is not taken as greater than 1.75 N/mm^2.	
Clause 3.8	shear span $a = 0.67$ m as before	
	$a/d = \dfrac{670}{285} = 2.4 < 6.0$	
	$f_v = \{[2.5 - 0.25(a/d)] \times 0.38\}$ $\quad = \{[2.5 - (0.25 \times 2.4)] \times 0.38\} = 0.72$ N/mm^2 $\qquad\qquad\qquad\qquad\qquad\qquad \le \ 1.75$ N/mm^2	
Table 8	Partial safety factor for shear strength $\gamma_{mv} = 2.0$	
	Design shear strength $= \dfrac{f_v}{\gamma_{mv}} = \dfrac{0.72}{2.0} = 0.36$ N/mm^2 $\qquad\qquad\qquad\qquad\qquad\qquad > \ 0.27$ N/mm^2	**Pocket wall is adequate with respect to shear**
Clause 8.2.4.3.1	**Note: Where the spacing of ribs exceeds 1.0 m, the ability of the masonry to span between the ribs should also be checked.**	

4.5   Prestressed Masonry

In prestressed masonry additional compressive forces are introduced to eliminate any tensile stresses induced by flexure. The principles of prestressing, which are well established and widely used in the concrete industry, are based on the algebraic addition of elastic bending and axial compressive stresses, i.e.

$$\text{Elastic bending stress} = \pm \frac{Bending\ moment}{Elastic\ section\ modulus} = \pm \frac{M}{Z}$$

$$\text{Elastic axial stress} = + \frac{Axial\ Load}{Cross\text{-}sectional\ area} = + \frac{P}{A}$$

The usual sign convention adopted when designing prestressed elements is **+ve** signifies **compression** and **−ve** signifies **tension**.
Note: *this is opposite to that normally adopted in structural analysis.*

The stresses can be due to dead loads, imposed loads and prestressing loads. The appropriate combinations which must be considered are dependent on the need to ensure that neither tensile stresses nor excessive compressive stresses will be induced under any given conditions as specified in the code, e.g. at the time of transfer of the load or in service.

   The principles are illustrated in the following example: consider a structural wall element with loading as shown in Figure 4.18.

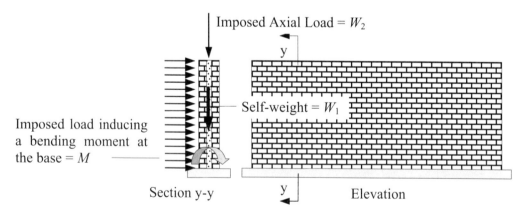

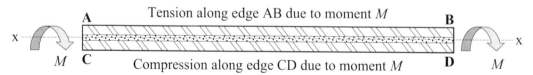

Plan

Figure 4.18

Axial stress in the wall due to the self-weight $\quad = +\dfrac{W_1}{A}$

Axial stress in the wall due to the imposed axial load $\quad = +\dfrac{W_2}{A}$

Flexural stress due to imposed load causing bending $\quad = \pm\dfrac{M}{Z}$

Combined stress along edge **AB** $\quad = \quad$ $+\dfrac{W_1}{A}+\dfrac{W_2}{A}-\dfrac{M}{Z}$ (1)

Combined stress along edge **CD** $\quad = \quad$ $+\dfrac{W_1}{A}+\dfrac{W_2}{A}+\dfrac{M}{Z}$ (2)

Clearly from equation (2) the maximum compressive stress occurs along edge **CD** and from equation (1) when $\left\{\dfrac{W_1}{A}+\dfrac{W_2}{A}\right\} < \dfrac{M}{Z}$ tension will occur along edge **AB**.

The introduction of a compressive prestressing force '**P**' can be used to eliminate the tension along **AB**. The force can be applied **concentric** to the centre-of-gravity of the cross-section producing a uniform stress or **eccentric** producing both an axial and a bending effect.

Consider an axial load '**P**' applied **concentric** to the centre-of-gravity:

Concentric prestressing force = **P**

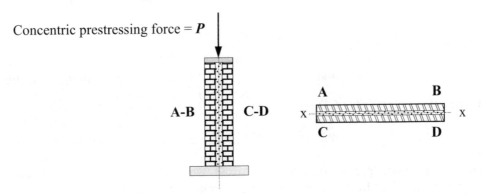

Figure 4.19

Axial stress in the wall due to the prestressing force **P** $\quad = +\dfrac{P}{A}$

To ensure that no tension occurs in the cross-section:

Combined stress along edge **AB** $= +\dfrac{W_1}{A}+\dfrac{W_2}{A}-\dfrac{M}{Z}+\dfrac{P}{A} \geq 0$

In addition the maximum compressive stress in the cross-section:

Combined stress along edge **CD** $= +\dfrac{W_1}{A}+\dfrac{W_2}{A}+\dfrac{M}{Z}+\dfrac{P}{A} \leq$ design compressive stress

Note: A **concentric** prestressing force is used when the cross-section can be subjected to a moment in either direction, e.g. due to wind loading on an external boundary wall exposed on both sides.

Consider an axial load '*P*' applied **eccentric** to the centre-of-gravity:

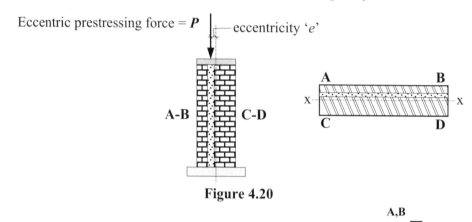

Figure 4.20

Axial stress in the wall due to the prestressing force $P = +\dfrac{P}{A}$

Flexural stress due to eccentricity '*e*' of force $P = \pm\dfrac{Pe}{Z}$

Combined stress along edge **AB** $= +\dfrac{W_1}{A} + \dfrac{W_2}{A} - \dfrac{M}{Z} + \left\{\dfrac{P}{A} + \dfrac{Pe}{Z}\right\} \geq 0$

and $\leq$ design compressive stress

Combined stress along edge **CD** $= +\dfrac{W_1}{A} + \dfrac{W_2}{A} + \dfrac{M}{Z} + \left\{\dfrac{P}{A} - \dfrac{Pe}{Z}\right\} \geq 0$

and $\leq$ design compressive stress

Note: An **eccentric** prestressing force is used when the cross-section can only be subjected to a moment from one direction, e.g. an earth retaining wall.

As indicated in Clause 9.1 of the code there are two methods which can be used to apply the required prestressing force, i.e.

♦ *Post-tensioning* in which tendons are tensioned against the masonry, when it has achieved sufficient strength, using mechanical anchorages, and

♦ *Pre-tensioning* in which tendons are tensioned against an independent anchorage and released only when the masonry and/or infill concrete has achieved sufficient strength. The transfer of the prestress force to the masonry is provided by bond alone.

Unless prefabricated products are being produced, post-tensioning is the most practical and normally adopted method in the UK. The detailing of any prestressing system used must ensure that the component parts are protected from corrosion and that the required prestress force is applied at the appropriate location with avoiding local overstressing of the masonry at the anchorage points. In Clause 9.1 the code clearly states: '*It is recommended that the tendons and anchorages for prestressed, post-tensioned masonry are inspectable unless corrosion resistant materials are used. A written statement for inspection and remedial action should be given to the owner.*'

A typical detail for applying a light prestressing force is shown in Figure 4.21(a) in which an anchorage nut at the top of a prestressing rod is tightened against a steel bearing plate through which the force is applied to the masonry. The magnitude of the prestressing force can be controlled by the use of a torque-wrench. It is important to provide a method for locking the nut in position after tightening. The prestressing rods can be anchored to the base as shown in Figures 4.21(b) and (c).

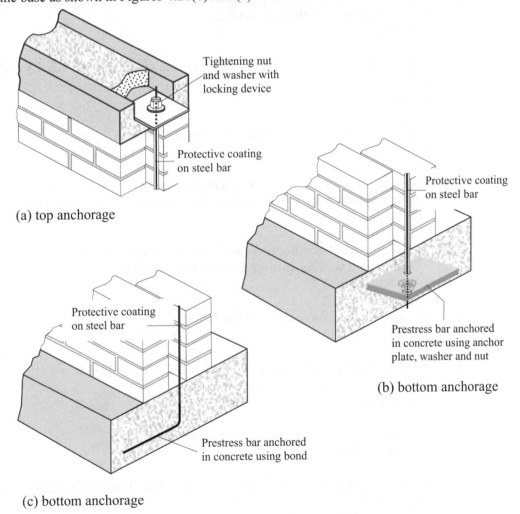

Tightening nut and washer with locking device

Protective coating on steel bar

(a) top anchorage

Protective coating on steel bar

Prestress bar anchored in concrete using anchor plate, washer and nut

(b) bottom anchorage

Protective coating on steel bar

Prestress bar anchored in concrete using bond

(c) bottom anchorage

Figure 4.21

4.6   Review Problems

4.1   Identify the categories of manufacturing and construction control which are permitted when designing reinforced/prestressed masonry.
(see section 4.2)

4.2   In which circumstances can mortar designation Type (iii) be used.
(see section 4.2)

4.3   Explain why the partial safety factors 'γ_m' are different from those used in unreinforced masonry.
(see section 4.2)

4.4   Identify five factors which influence the characteristic strength used in the design of reinforced flexural members.
(see section 4.2)

4.5   Explain the differences between the exposure conditions E1, E2, E3 and E4.
(see section 4.2)

4.6   Explain the term 'rectangular/parabolic stress block'.
(see section 4.2.1)

4.7   Explain the terms: 'balanced section', 'over-reinforced section' and 'under-reinforced section' indicating which one is preferable and why.
(see section 4.2.1)

4.8   Define the term 'deep beam' as used in the code and indicate how the area of reinforcing steel for such a beam can be determined.
(see section 4.2.1)

4.9   Define the term 'moment resistance factor'.
(see section 4.2.1)

4.10   State the additional design check which must be carried out in pocket-type walls where the rib spacing exceeds 1.0 m.
(see section 4.2.1)

4.11   Define the terms 'shear span' and 'principal load'.
(see section 4.2.2)

4.12   Explain why the provision of at least nominal links in a reinforced beam is advisable.
(see section 4.2.2)

4.13 Explain how the serviceability limit states of 'deflection' and 'cracking' are accommodated in the code.
(see section 4.2.3)

4.14 Explain how lateral torsional buckling is normally avoided when designing beams.
(see section 4.2.4)

4.15 Describe the circumstances under which the Table 9 values for deflection/cracking may be increased by 30%.
(see section 4.2.4)

4.16 Explain the purpose of providing a prestressing force in a masonry element.
(see section 4.5)

4.17 Explain the difference between a 'concentric' and an 'eccentric' prestress load and indicate where each type can be used.
(see section 4.5)

4.18 Explain the difference between pre-tensioning and post-tensioning.
(see section 4.5)

4.19 For which purpose does the code recommend that a written statement be given to the owner of a masonry structure.
(see section 4.5)

5. Overall Structural Stability

Objective: *'To introduce the concepts of structural stability, robustness and accidental damage.'*

5.1 Introduction

In the previous chapters the requirements of strength, stiffness and stability of individual structural components have been considered in detail. It is also **essential** in any structural design to consider the requirements of **overall** structural stability.

BS 5628 : Part 1 states the following in Clause 20.1:

'To ensure a robust and stable design it will be necessary to consider the layout of structure on plan, returns at the ends of walls, interaction between intersecting walls and the interaction between masonry walls and the other parts of the structure.

The design recommendations in section four (Part 1 of the code) assume that all the lateral forces acting on the whole structure are resisted by walls in planes parallel to these forces or by suitable bracing.

As well as the above general considerations, attention should be given to the following recommendations:

    a)   *buildings should be designed to be capable of resisting a uniformly distributed horizontal load equal to 1.5% of the total characteristic dead load above any level (see clause **22(b)** and (c));*

    b)   *connections of the type indicated in appendix C should be provided as appropriate at floors and roofs.'*

Appendix C gives illustrations of connections to floors and roofs by means of metal anchors and joist hangers capable of resisting lateral movement, (see Chapter 2, Section 2.1.5.3).

The term **stability** has been defined in *Stability of Buildings* published by the Institution of Structural Engineers (53) in the following manner:

'Provided that displacements induced by normal loads are acceptable, then a building may be said to be stable if:

    • *a minor change in its form, condition, normal loading or equipment would not cause partial or complete collapse and*

• *it is not unduly sensitive to change resulting from accidental or other actions.*
Normal loads include the permanent and variable actions for which the building has been designed.
The phrase "is not unduly sensitive to change" should be broadly interpreted to mean that the building should be so designed that it will not be damaged by accidental or other actions to an extent disproportionate to the magnitudes of the original causes of damage.'

This publication, and the inclusion of stability, robustness and accidental damage clauses in current design codes, is largely a consequence of the overall collapse or significant partial collapse of structures, e.g. the collapse of precast concrete buildings under erection at Aldershot in 1963 (54) and notably the Ronan Point Collapse due to a gas explosion in 1968 (55).

The Ronan Point failure occurred in May 1968 in a 23-storey precast building. A natural gas explosion in a kitchen triggered the progressive collapse of all of the units in one corner above and below the kitchen. The spectacular nature of the collapse had a major impact on the philosophy of structural design resulting in important revisions of design codes world-wide.

This case stands as one of the few landmark failures which have had a sustained impact on structural thinking.

The inclusion of such clauses in codes and building regulations is not new. The following is an extract from the 'CODE OF LAWS OF HAMMURABI (2200 BC), KING OF BABYLONIA' (the earliest building code yet discovered):

A. *If a builder builds a house for a man and do not make its construction firm and the house which he has built collapse and cause the death of the owner of the house - that builder shall be put to death.*

B. *If it cause the death of the son of the owner of the house - they shall put to death a son of that builder.*

C. *If it cause the death of a slave of the owner of the house - he shall give to the owner of the house a slave of equal value.*

D. *If it destroy property, he shall restore whatever it destroyed, and because he did not make the house which he built firm and it collapsed, he shall rebuild the house which collapsed at his own expense.*

E. *If a builder build a house for a man and do not make its construction meet the requirements and a wall fall in, that builder shall strengthen the wall at his own expense.*

Whilst this code is undoubtedly 'harsh' it probably did concentrate the designer's mind on the importance of structural stability!

An American structural engineer, Dr Jacob Feld, spent many years investigating structural failure and suggested ten basic rules to consider when designing and/or constructing any structure (44):

1. *Gravity always works, so if you don't provide permanent support, something will fail.*

2. *A chain reaction will make a small fault into a large failure, unless you can afford a fail-safe design, where residual support is available when one component fails. In the competitive private construction industry, such design procedure is beyond consideration.*

3. *It only requires a small error or oversight - in design, in detail in material strength, in assembly, or in protective measures - to cause a large failure.*

4. *Eternal vigilance is necessary to avoid small errors. If there are no capable crew or group leaders on the job and in the design office, then supervision must take over the chore of local control. Inspection service and construction management cannot be relied on as a secure substitute.*

5. *Just as a ship cannot be run by two captains, a construction job cannot be run by a committee. It must be run by one individual, with full authority to plan, direct, hire and fire, and full responsibility for production and safety.*

6. *Craftsmanship is needed on the part of the designer, the vendor, and the construction teams.*

7. *An unbuildable design is not buildable, and some recent attempts at producing striking architecture are approaching the limit of safe buildability, even with our most sophisticated equipment and techniques.*

8. *There is no foolproof design, there is no foolproof construction method, without guidance and proper and careful control.*

9. *The best way to generate a failure on your job is to disregard the lessons to be learnt from someone else's failures.*

10. *A little loving care can cure many ills. A little careful control of a job can avoid many accidents and failures.*

An appraisal of the overall stability of a complete structure during both the design and construction stages should be carried out by, and be the responsibility of, one individual. In many instances a number of engineers will be involved in designing various elements or sections of a structure but never the whole entity. It is **essential**, therefore, that one identified engineer carries out this vital appraisal function, including consideration of any temporary measures which may be required during the construction stage. In Clause 20.1 of the code it is clearly stated:

'*The designer responsible for the overall stability of the structure should ensure the compatibility of the design and details of parts and components. There should be no doubt of this responsibility for overall stability when some or all of the design and details are not made by the same designer.*'

5.2 Structural Form

Generally, instability problems arise due to an inadequate provision to resist lateral loading (e.g. wind loading) on a structure. There are a number of well-established structural forms which when used correctly, will ensure adequate stiffness, strength and stability. It is important to recognise that stiffness, strength and stability are three different characteristics of a structure. In simple terms:

- the *stiffness* determines the deflections which will be induced by the applied load system,
- the *strength* determines the maximum loads which can be applied before acceptable material stresses are exceeded and,
- the *stability* is an inherent property of the structural form which ensures that the building will remain stable.

The most common forms of structural arrangements which are used to transfer loads safely and maintain stability are:

- braced frames,
- unbraced frames,
- shear cores/walls,
- cross-wall construction,
- cellular construction,
- diaphragm action.

In many structures a combination of one or more of the above arrangements is employed to ensure adequate load paths, stability and resistance to lateral loading. All buildings behave as complex three-dimensional structures with components frequently interacting compositely to resist the applied force system. Analysis and design processes are a simplification of this behaviour in which it is usual to analyse and design in two dimensions with wind loading considered separately in two mutually perpendicular directions.

218   *Design of Structural Masonry*

5.3 Braced Frames

In braced frames lateral stability is provided in a structure by utilising systems of diagonal bracing in at least two vertical planes, preferably at right angles to each other. The bracing systems normally comprise a triangulated framework of members which are either in tension or compression. The horizontal floor or roof plane can be similarly braced at an appropriate level, as shown in Figure 5.1, or the floor/roof construction may be designed as a deep horizontal beam to transfer loads to the vertical, braced planes, as shown in Figure 5.2. There are a number of configurations of bracing which can be adopted to accommodate openings, services etc. and are suitable for providing the required load transfer and stability.

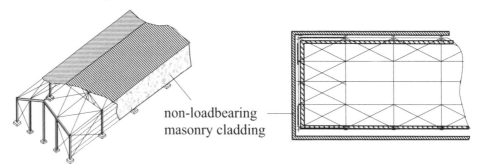

non-loadbearing masonry cladding

Figure 5.1   Braced frame

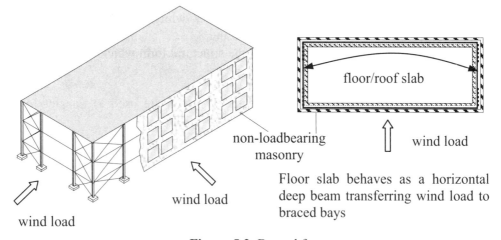

non-loadbearing masonry

wind load

floor/roof slab

wind load

wind load

Floor slab behaves as a horizontal deep beam transferring wind load to braced bays

Figure 5.2 Braced frames

In such systems the entire wind load on the building is transferred to the braced vertical planes and hence to the foundations at these locations.

5.4 Unbraced Frames

Unbraced frames comprise structures in which the lateral stiffness and stability are achieved by providing an adequate number of rigid (moment resisting) connections at appropriate locations. Unlike braced frames in which 'simple connections' only are required, the connections must be capable of transferring moments and shear forces. This is illustrated in the structure in Figure 5.3 in which stability is achieved in two mutually perpendicular directions using rigid connections. In wind direction A each typical transverse frame transfers its own share of the wind load to its own foundations through the moment connections and bending moments/shear forces/axial forces in the members. In wind direction B the wind load on either gable is transferred through the members and floors to stiffened bays (i.e. in the longitudinal section), and hence to the foundation at these locations. It is not necessary for *every* connection to be moment resisting.

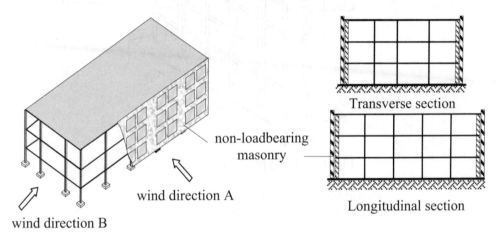

Figure 5.3 Unbraced frame

It is common for the portal frame action in a stiffened bay in wind direction B to be replaced by diagonal bracing whilst still maintaining the moment resisting frame action to transfer the wind loads in direction A.

As with braced frames, in most cases the masonry cladding and partition walls are non-loadbearing.

5.5 Shear Cores/Walls

The stability of modern high-rise buildings can be achieved using either braced or unbraced systems as described in Sections 5.3 and 5.4, or alternatively by the use of shear-cores and/or shear-walls. Such structures are generally considered as three-dimensional systems comprising horizontal floor plates and a number of strong-points provided by cores/walls enclosing stairs or lift shafts. A typical layout for such a building is shown in Figure 5.4.

In most cases the vertical loads are generally transferred to the foundations by a conventional skeleton of beams and columns whilst the wind loads are divided between several shear-core/wall elements according to their relative stiffness.

Where possible the plan arrangement of shear-cores and walls should be such that the centre-line of their combined stiffness is coincidental with the resultant of the applied wind load as shown in Figure 5.5.

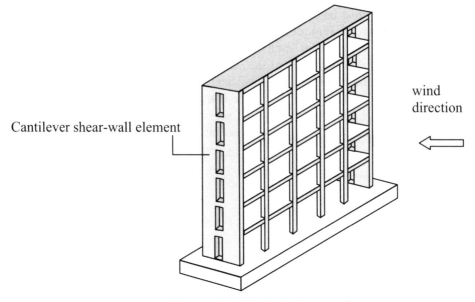

Cantilever shear-wall element

wind
direction

Figure 5.4   Typical shear-wall

centre-line of combined
stiffnesses of walls and core

wind
direction

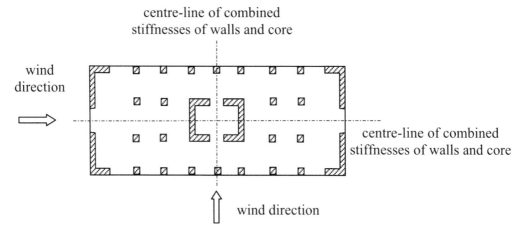

centre-line of combined
stiffnesses of walls and core

wind direction

Figure 5.5   Efficient layout of shear-core/walls

If this is not possible and the building is much stiffer at one end than the other, as in Figure 5.6, then torsion may be induced in the structure and must be considered. It is better at the planning stage to avoid this situation arising by selecting a judicious floor-plan layout. The floor construction must be designed to transfer the vertical loads (which are perpendicular to their plane) to the columns/wall elements in addition to the horizontal wind forces (in their own plane) to the shear-core/walls. In the horizontal plane they are designed as deep beams spanning between the strong-points.

There are many possible variations, including the use of concrete, steel, masonry and composite construction, which can be used to provide the necessary lateral stiffness, strength and stability.

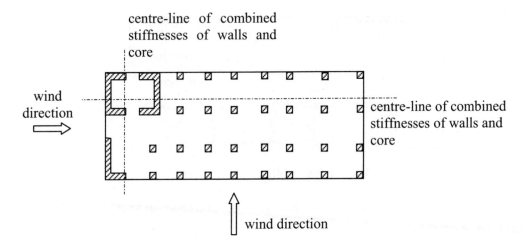

Figure 5.6   Inefficient layout of shear-core/walls

5.6   Cross-Wall Construction

In long rectangular buildings which have repetitive, compartmental floor plans such as hotel bedroom units and classroom blocks as shown in Figure 5.7, masonry cross-wall construction is often used.

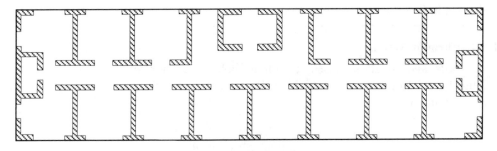

Figure 5.7   Cross-wall construction

Lateral stability parallel to the cross-walls is very high with the walls acting as separate vertical cantilevers sharing the wind load in proportion to their stiffnesses. Longitudinal stability, i.e. perpendicular to the plane of the walls, must be provided by the other elements such as the box sections surrounding the stair-wells/lift shafts, corridor and external walls.

5.7   Cellular Construction

It is common in masonry structures for the plan layout of walls to be irregular with a variety of exterior and interior walls as shown in Figure 5.8.

The resulting structural form is known as 'cellular construction', in which there is an inherent high degree of interaction between the intersecting walls. The provision of stair-wells and lift-shafts can also be integrated to contribute to the overall bracing of the structure.

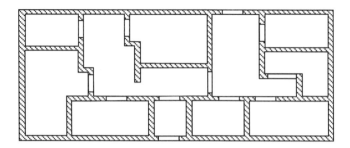

Figure 5.8 Cellular construction

It is important in both cross-wall and cellular masonry construction to ensure the inclusion of features such as:

♦   bonding or tying together of all intersecting walls,
♦   provision of returns where practicable at ends of load-bearing walls,
♦   provision of bracing walls to external walls,
♦   provision of internal bracing walls,
♦   provision of strapping of the floors and roof at their bearings to the load-bearing walls,

as indicated in *Stability of Buildings* (53).

5.8 Diaphragm Action

Floors, roofs and in some cases, cladding, behave as horizontal diaphragms which distribute lateral forces to the vertical wall elements. This form of structural action, is shown in Figure 5.9.

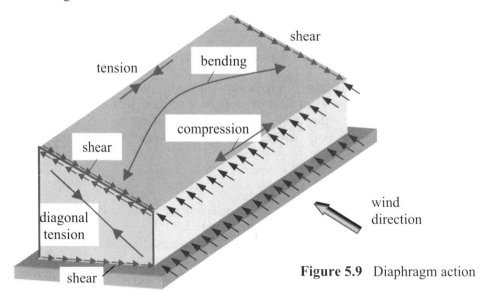

Figure 5.9 Diaphragm action

It is essential when utilising diaphragm action to ensure that each element and the connections between the various elements are capable of transferring the appropriate forces and providing adequate load-paths to the supports.

5.9 Accidental Damage and Robustness

It is inevitable that accidental loading such as vehicle impact or gas explosions will result in structural damage. A structure should be sufficiently robust to ensure that damage to small areas or failure of individual elements do not lead to progressive collapse or significant partial collapse. There are a number of strategies which can be adopted to achieve this, e.g.

- ◆ enhancement of continuity which includes increasing the resistance of connections between members and hence load transfer capability,
- ◆ enhancement of overall structural strength including connections and members,
- ◆ provision of multiple load paths to enable the load carried by any individual member to be transferred through adjacent elements in the event of local failure,
- ◆ the inclusion of load-shedding devices such as venting systems to allow the escape of gas following an explosion or specifically designed weak elements/details to prevent transmission of load.

The robustness required in a building may be achieved by 'tying' the elements of a structure together using peripheral and internal ties at each floor and roof level as indicated in   Figure 5.10.

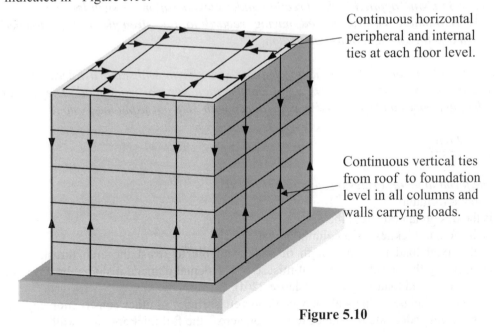

Continuous horizontal peripheral and internal ties at each floor level.

Continuous vertical ties from roof to foundation level in all columns and walls carrying loads.

Figure 5.10

An alternative to the 'fully tied' solution is one in which the consequences of the removal of each load-bearing member are considered in turn. If the removal of a member results in an unacceptable level of damage then this member must be strengthened to become a protected member (i.e. one which will remain intact after an accidental event), or the structural form must be improved to limit the extent of the predicted collapse. This process is carried out until all non-protected horizontal and vertical members have been removed one at a time.

In BS 5628 : Part 1 : 1992, Section 5 guidance is given to designers to limit the accidental damage and preserve the structural integrity of buildings with five storeys and above (i.e. Category 2 buildings). In Table 12 the code presents three options, any one of which should be considered by the designer, they are:

5.9.1 Option 1:

'*Vertical and horizontal elements, unless protected, proved removable, one at a time, without causing collapse.*'

Using this option the designer must examine the structural integrity of the building after the removal within each storey, bay, span or cantilever, of any single vertical or horizontal load-bearing element unless it is designed as a protected member.

A **protected member** is defined in the code in Clause 37.1.1 as '*a member in which, together with its essential supports, can withstand, without collapse, its reduced design load in accordance with Clause 22(d), and an accidental design load of 34 kN/m² applied from any direction together with the reaction, if any, which could be expected to be directly transmitted to that member by any attached building component also subjected to the load of 34 kN/m² applied in the direction under consideration, or such lesser reaction as might reasonably be transmitted having regard to the strength of the attached component and the strength of its connection.*

A masonry column or wall may have adequate strength to withstand a lateral design pressure of 34 kN/m² if it supports a sufficiently high vertical axial load. The lateral strength of masonry can be checked in accordance with **36.8** *but using the formula*

$$q_{lat} = \frac{7.6\ tn}{h_a^2}$$

Note the formula incorporates a safety factor of 1.05.'
In the above formula:

q_{lat} is the design lateral strength,
t is the actual thickness of a column or wall,
n is the axial load per unit length of wall available to resist the arch thrust; when considering the possible effects of misuse or accidental damage it should be based on the design dead load as given in Clause 22(d),
h_a is the clear height of the wall between concrete surfaces or other construction capable of providing adequate resistance to rotation across the full thickness of a wall.

The definitions of load-bearing elements (i.e. beams, columns, slab/floor/roof construction and walls) for the purposes of considering accidental damage are given in Table 11 of the code.

5.9.2 *Option 2:*

'*Horizontal ties; Peripheral, internal and column or wall in accordance with 37.3 and table 13,*
Vertical ties; None or ineffective.'
Using this option, horizontal ties are provided at each floor and roof level (with the exception of lightweight roofs, i.e. *roofs comprising timber or steel trusses, flat timber roofs or roofs incorporating concrete or steel purlins with asbestos or wood-wool deck*) as indicated in Table 13 of the code. These requirements are illustrated in Figure 5.11 to 5.17.
The magnitude of the basic tie force is given by:

$$F_t \leq 60 \text{ kN}$$
$$\leq (20 + 4N_s) \text{ kN} \quad \text{(or kN/m in the case of internal and external wall ties)}.$$

This value is modified according to the type of tie being considered.
There are four types of horizontal tie:

A. *Peripheral ties:*

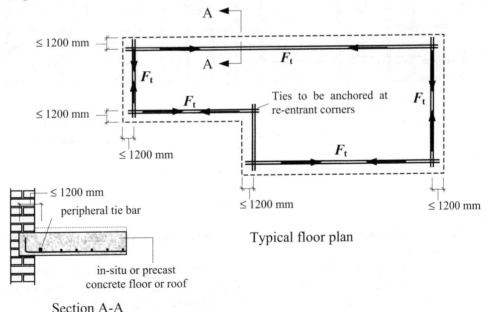

Typical floor plan

Section A-A

Figure 5.11

The anchoring of ties is normally achieved by providing an adequate length of bar based on the bond strength of the concrete.

B. *Internal ties:*

Internal ties are provided in two orthogonal directions, parallel to and perpendicular to, the direction of the span of the member being considered, as shown in Figures 5.12 to 5.15 for one-way spanning and two-way spanning slabs respectively.

The ties should be anchored to perimeter ties or continue as wall or column ties. The location and distribution of the ties can take one of four forms:

(i)   provided uniformly throughout the floor or roof width,

(ii)   concentrated at locations not more than 6.0 m apart, e.g. coinciding with the ribs positions in T-beam floor construction (see Figure 5.13),

(iii) located within walls no more than 0.5 m above or below the floor or roof level and not more than 6.0 m apart in a horizontal direction (see Figure 5.14),

(iv)   placed evenly in the perimeter zone (in addition to the peripheral ties).

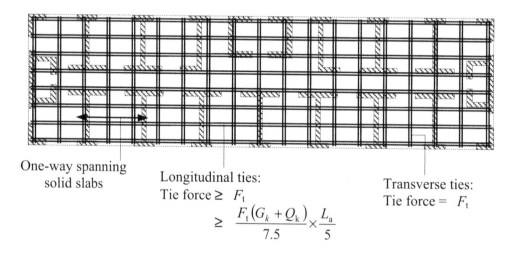

One-way spanning
solid slabs

Longitudinal ties:
Tie force $\geq F_t$

$$\geq \frac{F_t(G_k + Q_k)}{7.5} \times \frac{L_a}{5}$$

Transverse ties:
Tie force $= F_t$

Figure 5.12

where:

$(G_k + Q_k)$ is the sum of the average characteristic dead and imposed loads in kN/m^2 ,
L_a is the lesser of:

the greatest distance in metres in the direction of the tie, between the centres of columns or other load-bearing members whether this distance is spanned by a single slab or by a system of beams and slabs   or
$5 \times$ the clear storey height '*h*'

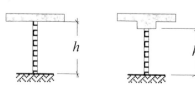

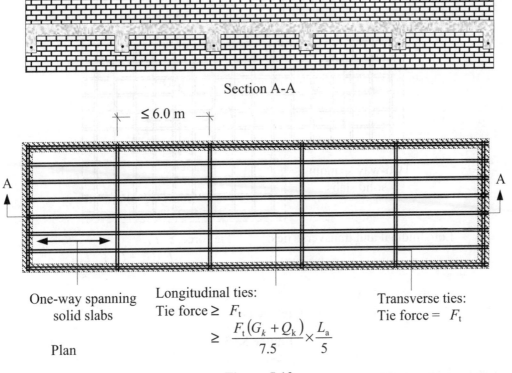

Section A-A

≤ 6.0 m

A

A

One-way spanning
solid slabs

Plan

Longitudinal ties:
Tie force ≥ F_t

$$\geq \frac{F_t(G_k + Q_k)}{7.5} \times \frac{L_a}{5}$$

Transverse ties:
Tie force = F_t

Figure 5.13

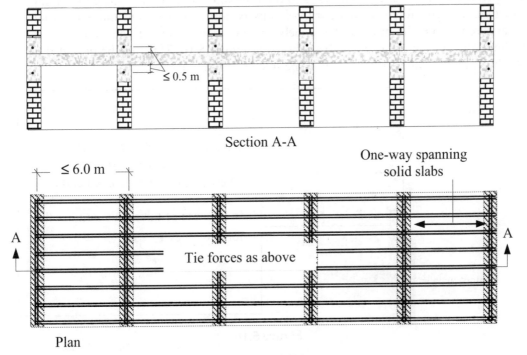

≤ 0.5 m

Section A-A

One-way spanning
solid slabs

≤ 6.0 m

A

A

Tie forces as above

Plan

Figure 5.14

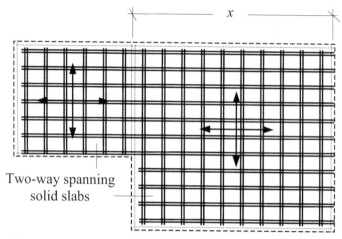

Longitudinal and transverse ties: Tie force $\geq F_t$

$$\geq \frac{F_t(G_k + Q_k)}{7.5} \times \frac{L_a}{5}$$

where $L_a \leq x$

              $\leq 5 \times$ clear storey height

Figure 5.15

C. External column ties:

The purpose of external column ties is to hold the top and bottom of the column into the structure. The 'tie force' in masonry columns is normally achieved through **either** the shear strength **or** the friction force between the bottom/top of the column and the top/underside of the floor/roof slab as shown in Figure 5.16. This may be provided partly or wholly by the same reinforcement as perimeter and internal ties.

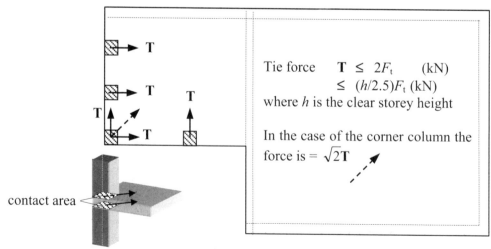

Tie force $T \leq 2F_t$ (kN)
                    $\leq (h/2.5)F_t$ (kN)
where h is the clear storey height

In the case of the corner column the force is $= \sqrt{2}T$

Figure 5.16

The characteristic shear strength is given in Clause 25 as $f_v = 0.35$ N/mm^2 (assuming a minimum value with g_A = zero). In Clause 27.4 the partial safety factor for shear loads 'γ_{mv}' is given as 1.25 when considering the probable effects of misuse or accident.

*Considering the **corner column** where the tie force is resisted by shear:*

Design shear strength = $\dfrac{f_v}{\gamma_{mv}} \quad \dfrac{0.35}{1.25}$ = 0.28 N/mm^2

$\therefore$ (0.28 × contact area) $\geq \sqrt{2}\,\text{T}$ $\quad \therefore$ minimum contact area = $\dfrac{\sqrt{2}\text{T}}{0.28}$ = **5.05T**

In most cases there are two shear planes, which resist the tie force as shown in Figure 5.16, and the contact area is equal to **2.53T**.

*Considering the **corner column** where the tie force is resisted by friction:*

Design friction resistance = $\dfrac{(coefficient\ of\ friction)\times(least\ favourable\ vertical\ load)}{(overall\ factor\ of\ safety)}$

$\qquad = \dfrac{0.6\times n}{1.05}$ = 0.57n

where:
coefficient of friction $\qquad\qquad$ = 0.6 as given in Clause 26
n is the least favourable vertical load = 0.95 G_k as given in Clause 22 (d)
overall factor of safety $\qquad\qquad$ = 1.05 as indicated in Clause 37.1.1

$\therefore$ (0.57n × contact area) $\geq \sqrt{2}\,\text{T}$ $\quad \therefore$ minimum contact area = $\dfrac{\sqrt{2}\text{T}}{0.57n}$ = **2.48$\dfrac{\text{T}}{n}$**

As before in most cases there are two shear planes which resist the tie force as shown in Figure 5.16 and the contact area is equal to **1.24$\dfrac{\text{T}}{n}$**.

D. *External wall ties:*
As with columns the purpose of external wall ties is to hold the top and bottom of the wall into the structure and the 'tie force' is normally achieved through **either** the shear strength **or** the friction force. This may be provided partly or wholly by the same reinforcement as perimeter and internal ties. In the case of walls where ties are required they should be:

(i) spaced uniformly along the length of the wall (as shown in Figure 5.17),or

(ii) concentrated at centres not more than 5.0 m apart and not more than 2.5 m from the end of the wall.

The minimum contact between the walls and the slab surfaces required using either shear or friction strength is calculated in a similar manner to that for columns. It is possible to provide ties concentrated at locations along the wall by the use of pockets etc., however, it does present additional detailing difficulties and the most practical solution is to ensure adequate shear or friction strength is available.

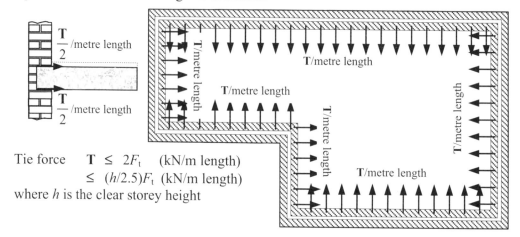

Tie force $\quad \mathbf{T} \leq 2F_t \quad$ (kN/m length)
$\quad\quad\quad\quad\quad \leq (h/2.5)F_t$ (kN/m length)
where h is the clear storey height

Figure 5.17

In summary, Option 2 requires the provision of horizontal peripheral and internal ties in addition to column and wall ties at their junction with the floor slabs. Vertical ties extending from roof to foundation level are not required but vertical members must either be '*protected members*' or alternatively structural robustness must be checked after their removal.

5.9.3 Option 3:

'*Horizontal ties; Peripheral, internal and column or wall in accordance with 37.3 and table 13,*
Vertical ties; In accordance with 37.4 and table 14 (see Figure 5.10).'

Using this option, horizontal ties are provided as in Option 2 and in addition vertical ties are provided to satisfy the following requirements:
    Precast or insitu concrete or other heavy floor or roof units should be anchored, in the direction of their span, either to each other over a support or directly to their supports, in such a manner as to be capable of resisting a horizontal tensile force of F_t kN/metre width, where F_t is as given in Table 13. The wall should be contained between concrete surfaces or other similar construction, excluding timber, capable of providing resistance to lateral movement and rotation across the full width of the wall.
    The vertical ties should extend from the roof level to the foundation or to a level at and below which the relevant members of the structure are protected in accordance with Clause 37.1.1. They should also be fully anchored at each end and at each floor level and any joint should be capable of transmitting the required tensile forces.
    No additional calculations are required to consider the effect of removing individual elements when using Option 3.
    Although this option provides rules to incorporate robustness within a structure they should not be followed 'blindly' and the designer must give careful consideration to the

provision of ties and the detailing required to enable them to fulfil their intended purpose, e.g. protection against corrosion and damage of the ties. When using Option 3 the physical problems associated in locating the ties makes it unattractive to designers and it is more likely that one of the other options will be adopted in most cases. More detailed and comprehensive information relating to the provision of ties can be found in references (30 and 46).

Note: In all cases the required area of steel for ties is **not** in addition to that already provided and serving other purposes, e.g. reinforcement in concrete or masonry required to resist tension under normal loading

The structural integrity of the building is then examined after the removal of any single vertical loadbearing element as in Option 1.

5.10 Repair and Maintenance

The need to rehabilitate and restore old and damaged buildings may arise because of their architectural or historical significance or simply be due to economic need, e.g. during a rapidly expanding period of urban regeneration. Any repair and maintenance programme must be undertaken in a rational and methodical manner including:

(i) conducting a detailed survey to identify the nature and extent of any defects which exist,

(ii) carrying out an analysis of the data obtained from the survey (including mathematical modelling if necessary), to determine the cause(s) of the problems,

(iii) formulating a program of work to undertake the most appropriate remedial measures required, and

(iv) ensuring that the individual(s) contracted to undertake the work are knowledgeable and experienced in the type of work involved, which can often be of a specialist nature.

The need for a detailed survey supported by mathematical analysis where appropriate was recognised as early as the mid-18th. century when the dome of St Peter's Basilica in Rome was strengthened by the addition of five rings to the cupola after much debate and the more traditional approach of '*empirical rules and opinion*' was discarded.

The main defects in masonry are caused by problems such as overloading, differential settlement of the foundations, movement of the ground, water penetration, removal of load-bearing walls to accommodate openings and corrosion of built-in components. All of these can induce varying degrees of leaning, bowing, cracking or crushing of both the masonry and any softer materials such as timber imbedded in the masonry, providing visual clues to the cause(s) of the problems.

A comprehensive treatment of this topic is beyond the scope of this text and a number of publications have been produced which provide excellent reference material for engineers planning to undertake any appraisal and repair of existing masonry, i.e. (47, 51, 57) and readers are recommended to refer to these for further information.

5.11 Review Problems

5.1   Distinguish between the strength, stiffness and stability of a building.
       (see section 5.2)

5.2   Identify five methods of providing lateral stability to a building.
       (see section 5.2)

5.3   Identify five features which are important for stability in buildings of
       cellular construction.
       (see section 5.7)

5.4   Explain the importance of connections in diaphragm action.
       (see section 5.7)

5.5   Explain the purpose of horizontal and vertical ties in a building.
       (see section 5.8)

5.6   Explain the differences between Options 1, 2 and 3 when designing to
       resist accidental damage.

6. Eurocode 6

> **Objective:** *'To provide an introduction to Eurocode 6, the limit state design code for the design of structural masonry elements.'*

6.1 Introduction

This chapter provides an introduction to the contents of the Eurocode 6 : Part 1.1. The intention is to produce EC 6 in a number of parts, the provisional titles of which are:

EC 6: Part 1.1
'General rules for buildings – rules for reinforced and unreinforced masonry'

EC 6: Part 1.2
'General rules – Structural fire design'

EC 6: Part 1.3
'General rules – Detailed rules on lateral loading'

EC 6: Part 1.X
'Complex shape sections in masonry structures'

EC 6: Part 2
'Design, selection of materials and execution of masonry'

EC 6: Part 3
'Simplified and simple rules for masonry structures'

EC 6: Part 1.1
'Constructions with lesser requirements for reliability and durability'

Currently only Parts 1.1 and 1.2 are available as DD ENV 1996-1-1 : 1996 and DD ENV 1996-1-2 :1997 which were published in 1996 and 1997 respectively. The remaining section are under consideration.

The European Standards Organisation, CEN, is the umbrella organisation under which a set of common structural design standards (e.g. EC1, EC2, EC3, etc.) has been developed. The Structural Eurocodes are the result of attempts to eliminate barriers to trade throughout the European Union. Separate codes exist for each structural material, including EC6: Part 1.1 for masonry. The basis of design and loading considerations are included in EC1.

Each country publishes its own European Standards (EN), e.g. in the UK the British Standards Institution (BSI) issues documents (which are based on the Eurocodes developed under CEN), with the designation BS EN.

Structural Eurocodes are currently issued as Pre-standards (ENV) which can be used as an alternative to existing national rules. In the UK the BSI has used the designation DD ENV; the pre-standards are equivalent to the traditional 'Draft for development' Documents.

In the UK the Eurocode for masonry design is known as 'DD ENV 1996-1-1 Eurocode 6 : Design of masonry structures : Part 1-1 General rules and rules for buildings – Rules for reinforced and unreinforced masonry (together with the United Kingdom National Application Document)'.

Eurocode 6 adopts the 'Limit State Design' philosophy as currently used in UK national standards.

6.1.1   *National Application Document (NAD)*

Each country which issues a European Standard also issues a NAD for use with the EN. The purpose of the NAD is to provide information to designers relating to product standards for materials, partial safety factors and any additional rules and/or supplementary information specific to design within that country.

A summary of the abbreviations used above is given in Table 6.1.

6.2   Terminology, Symbols and Conventions

The terminology, symbols and conventions used in EC6: Part 1.1 differ from those used by BS 5628 : Part 1 : 1992. The code indicates **'Principles'** which are general statements and definitions which must be satisfied and **'Rules'** which are design procedures which comply with the Principles. The rules can be substituted by alternative procedures provided that these can be shown to be in accordance with the Principles.

There are two types of Annexe in EC6: Part 1.1: normative and informative. Normative Annexes have the same status as the main body of the text whilst Informative Annexes provide additional information. The Annexes generally contain more detailed material or material which is used less frequently.

6.2.1   *Decimal Point*

Standard ISO practice has been adopted in representing a decimal point by a comma, i.e. $5,3 \equiv 5.3$.

Abbreviation	Meaning
CEN	European Standards Organisation
EC	Eurocode produced by CEN
EN	European Standard based on Eurocode and issued by member countries
ENV	Pre-standard of Eurocode issued by member countries
DD ENV	UK version of Pre-standard (BSI)
NAD	National Application Document issued by member countries (BSI)

Table 6.1

6.2.2   Symbols and Subscripts

As in BS 5628 there are numerous symbols[*] and subscripts used in the code. There are in excess of 150 variables defined in the code relating to loading, partial safety factors, loading effects and material properties. They are too numerous to identify and some of the most frequently used ones are given here for illustration purposes:

Symbols:

F: *action*,   a force (load) applied to a structure or an imposed deformation (indirect action), such as temperature effects or settlement

G: permanent action such as dead loads due to self-weight, e.g.

Characteristic value of a permanent action $= G_k$
Design value of a permanent action $= G_d$
Lower design value of a permanent action $= G_{d,inf}$
Upper design value of a permanent action $= G_{d,sup}$

Q: variable actions such as imposed, wind or snow loads,

Characteristic value of a variable action $= Q_k$
Design value of a variable action $= Q_d$

A: accidental actions such as explosions, fire or vehicle impact.

E: *effect of actions* on static equilibrium or of gross displacements etc. e.g.

Design effect of a destabilising action $= E_{d,dst}$

Design effect of a stabilising action $= E_{d,stb}$

Note: $E_{d,dst} \leq E_{d,stb}$

R: *design resistance* of structural elements, e.g.

Design vertical load resistance of a wall $= N_{Rd} = \dfrac{\Phi_{i,m} t f_k}{\gamma_M}$

Design shear resistance of a wall $= V_{Rd} = \dfrac{f_{vk} t l_c}{\gamma_M}$

Design moment of resistance of a section $= M_{Rd} = \dfrac{A_s f_{yk} z}{\gamma_s}$

---

[*] In most cases the Eurocode does not use italics for variables.

S: *design value of actions*   factored values of externally applied loads or load effects such as axial load, shear force, bending moment etc. e.g.

Design vertical load	$N_{Sd} \leq N_{Rd}$
Design shear force	$V_{Sd} \leq V_{Rd}$
Design bending moment	$M_{Sd} \leq M_{Rd}$

X: *material property* physical properties such as tension, compression, shear and bending strength, modulus of elasticity etc. e.g.

Characteristic compressive strength of masonry $X_k = f_k = K f_b^{0,65}$ N/mm²

Design compressive strength of masonry $= X_d = \dfrac{X_k}{\gamma_M} = f_d = \dfrac{K f_b^{0,65}}{\gamma_M}$ N/mm²

6.3 Limit State Design

The limit states are states beyond which a structure can no longer satisfy the design performance requirements (see Chapter 1, section 1.5). The two classes of limit state adopted by EC6: Part 1.1 are:

♦ *ultimate limit states:*   These include failures such as full or partial collapse due to e.g. rupture of materials, excessive deformations, loss of equilibrium or development of mechanisms. Limit states of this type present a direct risk to the safety of individuals.

♦ *serviceability limit states:*   Whilst not resulting in a direct risk to the safety of people, serviceability limit states still render the structure unsuitable for its intended purpose. They include failures such as excessive deformation resulting in unacceptable appearance or non-structural damage, loss of durability or excessive vibration causing discomfort to the occupants.

The limit states are quantified in terms of design values for actions, material properties and geometric characteristics in any particular design. Essentially the following conditions must be satisfied:

Ultimate limit state:

Rupture $S_d \leq R_d$

where:

S_d is the design value of the effects of the actions imposed on the structure/structural elements,

R_d is the design resistance of the structure/structural elements to the imposed actions.

Stability $S_{d,dst} \leq R_{d,stb}$

where:

$S_{d,dst}$ is the design value of the destabilising effects of the actions imposed on the structure (including self-weight where appropriate).

$S_{d,stb}$ is the design value of the stabilising effects of the actions imposed on the structure (including self-weight where appropriate).

Serviceability limit state:

$$\textbf{Serviceability} \quad \textbf{S}_\textbf{d} \leq \textbf{C}_\textbf{d}$$

where:

S_d is the design value of the effects of the actions imposed on the structure/structural elements,

C_d is a prescribed value, e.g. a limit of deflection.

6.3.1 Design Values

The term *design* is used for factored loading and member resistance

Design loading $(F_d) = $ partial safety factor $(\gamma_F) \times$ characteristic value (F_k)

e.g. $$G_d = \gamma_G G_k$$

where:

γ_G is the partial safety factor for permanent actions,

G_k is the characteristic value of the permanent actions.

Note: $G_{d,sup}$ $(= \gamma_{G,sup}G_{k,sup}$ or $\gamma_{G,sup}G_k)$ represents the 'upper' design value of a permanent action,

$G_{d,inf}$ $(= \gamma_{G,inf}G_{k,inf}$ or $\gamma_{G,inf}G_k)$ represents the 'lower' design value of a permanent action.

$$\text{Design resistance} \quad (R_d) \quad = \quad \frac{\text{material characteristic strength} \,(X_k)}{\text{material partial safety factor} \,(\gamma_m)}$$

e.g. ***Design flexural strength*** $=$ $f_{x,d}$ $=$ $\dfrac{f_{x,k}}{\gamma_m}$

The design values of the actions vary depending upon the limit state being considered. All of the possible load cases should be considered in different combinations as given in Clause 2.3.2.2 (2)P and Table 2.1 of the code, and in Tables 1 and 2 of the NAD; e.g. for persistent and transient design situations:

$$F_d \; = \; \Sigma\gamma_{G,j}\,G_{k,j} + \gamma_{Q,1}Q_{k,1} + \underset{i>1}{\Sigma\gamma_{Q,i}}\,\psi_{0,i}\,Q_{k,i} \qquad \begin{array}{l}\textbf{Equation (1)}\\ \text{(Equation (2.17) in EC6)}\end{array}$$

where:

$\gamma_{G,j}$ partial safety factor for permanent actions, (Table 1 of the NAD)

$G_{k,j}$ characteristic values of permanent actions,

$\gamma_{Q,1}$ partial safety factor for *'one'* of the variable actions, (Table 1 of the NAD)

$Q_{k,1}$ characteristic value of *'one'* of the variable actions,

$\gamma_{Q,i}$ partial safety factor for the other variable actions, (Table 1 of the NAD)

$\psi_{0,i}$ combination factor which is applied to the characteristic value Q_k of an action not being considered as $Q_{k,1}$, (Eurocode 1 and NAD - Table 2)

$Q_{k,i}$ characteristic value of *'other'* variable actions.

Table 2.1 from the code and Tables 1 and 2 from the NAD are given in Figures 6.1, 6.2 and 6.3 respectively.

Extract from EC6: Table 2.1

Design values for actions for use in the combination of actions				
Design situation	Permanent actions G_d	Variable actions		Accidental actions A_d
		One with its characteristic value	Others with their combination value	
Persistent and Transient	$\gamma_G G_k$	$\gamma_Q Q_k$	$\psi_0 \gamma_Q Q_k$	–
Accidental	$\gamma_{GA} G_k$	$\psi_1 Q_k$	$\psi_2 Q_k$	$\gamma_A A_k$ (if A_d is not specified directly)

Figure 6.1

Extract from NAD Table 1: (full table not given)

Table 1. Partial safety factors (γ factors)					
Reference in ENV 1996-1-1	Definition	Symbol	Condition	Value	
				Boxed ENV 1996-1-1	UK
2.3.3.1	Partial factors for variable actions	γ_A	Accidental	1,0	1,0
		$\gamma_{F,inf}$	Favourable	0,0	0,0
		γ_Q	Unfavourable	1,5	1,5
		γ_Q	Reduced Favourable	0,0	0,0
		γ_Q	Reduced Unfavourable	1,35	1,35
2.3.3.1	Partial factors for permanent actions	γ_{GA}	Accidental	1,0	1,0
		γ_G	Favourable	1,0	1,0
		γ_G	Unfavourable	1,35	1,35
		$\gamma_{G,inf}$	Favourable	0,9	0,9
		$\gamma_{G,sup}$	Unfavourable	1,1	1,1
		γ_p	Favourable	0,9	0,9
		γ_p	Unfavourable	1,2	1,2

Figure 6.2

Extract from NAD Table 2:

Table 2. combination factors (ψ factors)				
Variable action	**Building type**	ψ_0	ψ_1	ψ_2
Imposed floor loads	Dwellings	0,5	0,4	0,2
	Other occupancy classes[1]	0,7	0,6	0,3
	Parking	0,7	0,7	0,6
Imposed roof loads	All occupancy classes[1]	0,7	0,2	0,0
Wind loads	All occupancy classes[1]	0,7	0,2	0,0
[1] As listed and defined in table 1 of BS 6399 : Part 1 : 1984				

Figure 6.3

6.3.2 Partial Safety Factors

The Eurocode provides indicative values for various safety factors: these are shown in the text as 'boxed values' e.g. $\boxed{1,35}$. Each country defines 'boxed values' within the NAD document to reflect the levels of safety required by the appropriate authority of the national government; in the UK i.e the British Standards Institution.

The boxed values of partial safety factors for actions in building structures for persistent and transient design situations are given in Table 2.2 of EC6: Part 1.1 and in Figure 6.4 of this text (these values are also given in Table 1 of the NAD).

Extract from EC6:Table 2.2

Table 2.2 Partial safety factors for actions in building structures for persistent and transient design situations.				
	Permanent actions (γ_G) (see note)	Variable actions (γ_Q)		Prestressing (γ_p)
		One with its characteristic value	Others with their combination value	
Favourable effect	$\boxed{1,0}$	$\boxed{0}$	$\boxed{0}$	$\boxed{0,9}$
Unfavourable effect	$\boxed{1,35}$	$\boxed{1,5}$	$\boxed{1,35}$	$\boxed{1,2}$
Note: See also paragraph 2.3.3.1 (3) – {see section 6.3.2 of this text}.				

Figure 6.4

Consider a design situation in which there are two characteristic dead loads, G_1 and G_2, in addition to three characteristic imposed loads, Q_1, Q_2 and Q_3. Assume the partial safety factors and combination factor are $\gamma_{G,j} = 1,35$, $\gamma_{Q,1} = 1,5$ $\gamma_{Q,i} = 1,5$ $\psi_{0,i} = 0,7$.

Combination 1: $F_d = (1,35G_1 + 1,35G_2) + 1,5\,Q_1 + (1,5 \times 0,7 \times Q_2) + (1,5 \times 0,7 \times Q_3)$

$\qquad\qquad\quad \mathbf{F_d = 1,35(G_1 + G_2) + 1,5Q_1 + 1,05(Q_2 + Q_3)}$

Combination 2: $\mathbf{F_d =}$ $\mathbf{1{,}35(G_1 + G_2) + 1{,}5Q_2 + 1{,}05(Q_1 + Q_3)}$
Combination 3: $\mathbf{F_d =}$ $\mathbf{1{,}35(G_1 + G_2) + 1{,}5Q_3 + 1{,}05(Q_1 + Q_2)}$

When developing these combinations permanent effects are represented by their upper design values, i.e.

$$G_{d,sup} = \gamma_{G,sup}\, G_{k,sup} \quad \text{or} \quad \gamma_{G,sup}\, G_k$$

Those which decrease the effect of the variable actions (i.e. favourable effect) are replaced by their lower design values, i.e.

$$G_{d,inf} = \gamma_{G,inf}\, G_{k,inf} \quad \text{or} \quad \gamma_{G,inf}\, G_k$$

In most situations either the upper or lower design values are applied throughout the structure; specifically in the case of continuous beams, the same design value of self-weight is applied on all spans.
    A similar approach is used when dealing with accidental actions.

In Clause 2.3.3.1(3)P two simplified expressions using the Table 2.2 values are given to replace Equation (1). They are:
considering the most unfavourable variable action

$$F_d = \Sigma\gamma_{G,j}\, G_{k,j} + 1{,}5\, Q_{k,1} \qquad \textbf{Equation (2)}$$
$$\text{(Equation (2.19) in EC6)}$$

considering all unfavourable variable actions

$$F_d = \Sigma\gamma_{G,j}\, G_{k,j} + 1{,}35\, \Sigma Q_{k,i} \qquad \textbf{Equation (3)}$$
$$\underset{i > 1}{} \qquad \text{(Equation (2.20) in EC6)}$$

whichever gives the larger value.

6.4 Conventions

The difference in conventions most likely to cause confusion with UK engineers is the change in the symbols used to designate the major and minor axes of a cross-section. Traditionally in the UK the **y-y axis** has represented the minor axis; in EC6 this represents the **MAJOR axis,** the minor axis is represented by the z-z axis. The **x-x axis** defines the **LONGITUDINAL axis**. All three axes are shown in Figure 6.5.

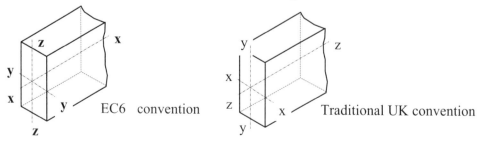

EC6 convention   Traditional UK convention

Figure 6.5

6.5 Materials

In Chapter 3 of the code information is given regarding the physical properties of materials, i.e. masonry units, mortar, concrete infill, reinforcing steel, prestressing steel and ancillary components such as damp-proof courses, wall ties etc. The most frequently used materials relate to masonry units and mortar both of which are discussed in this text, the reader is referred to the code for further information relating to the other materials.

6.5.1 *Masonry Units*

The masonry units considered in the code are those manufactured from:

Clay	(EN 771*-1),
Calcium silicate	(EN 771-2),
Aggregate concrete (dense and lightweight aggregate)	(EN 771-3),
Autoclaved aerated concrete (a.a.c)	(EN 771-4),
Manufactured stone	(EN 771-5),
Dimensioned natural stone	(EN 771-6).

* Note: EN 771 is the proposed code for masonry product and testing standards which has still to be completed and published.

The masonry units are classified into four Groups: 1, 2a, 2b and 3 in relation to the percentage of voids/holes existing in each unit. The most solid units are in Group 1 whilst the units in Group 3 generally have the highest percentage of voids. This classification is given in Table 3.1 of the code; see Figure 6.6.

In the UK, all of the brick manufacturers' and most of the concrete block manufacturers' products correspond to Group 1. There are some concrete products which correspond to Group 2a. The classification into Groups is used when determining the characteristic compressive and shear strength of unreinforced masonry.

6.5.1.1 *Characteristic Compressive Strength of Unreinforced Masonry* (f_k)

The characteristic compressive strength of unreinforced masonry is defined in Clause 3.6.2 as:

$$f_k = K f_b^{0,65} f_m^{2,25} \text{ N/mm}^2 \quad \text{when using '\textbf{\textit{general purpose}}' mortar,}$$

$$f_k = 0,8 f_b^{0,65} \text{ N/mm}^2 \quad \text{when using '\textbf{\textit{thin layer}}' mortar,}$$

$$f_k = K f_b^{0,65} \text{ N/mm}^2 \quad \text{when using '\textbf{\textit{lightweight}}' mortar.}$$

where:
f_k is the characteristic compressive strength of mortar,
f_b is the normalised compressive strength of masonry.
f_m is the compressive strength of mortar,
K is a constant.

The normalised compressive strength is defined in Clause 3.1.2.1 of the code:

♦ '*When the compressive strength of masonry units is quoted as the **mean** strength when tested in accordance with EN 772-1 this should be converted to the normalised compressive strength by converting to the air dried strength, if it is not*

Extract from EC 6: Table 3.1

	Group of Masonry Units			
	1	**2a**	**2b**	**3**
Volume of holes (% of the gross volume) (see note 1)	≤ 25	> 25 - 45 for clay units > 25 - 50 for concrete aggregate units	> 45 - 55 for clay units > 50 - 60 for concrete aggregate units (see note 2)	≤ 70
Volume of any hole (% of the gross volume)	< 12,5	$\leq$ 12,5 for clay units $\leq$ 25 for concrete aggregate units	$\leq$ 12,5 for clay units < 25 for concrete aggregate units	Limited by area (see below)
Area of any hole	Limited by area (see above)	Limited by area (see above)	Limited by area (see above)	$\leq 2\,800$ mm^2 except for units with a single hole when the hole should be $\leq 18\,000$mm^2
Combined thickness (% of the overall width) (see note 3)	$\geq 37,5$	≥ 30	≥ 20	No requirement

Notes:

1. Holes may consist of formed vertical holes through the units or frogs or recesses.

2. If there is national experience, based on tests, that confirms that the safety of the masonry is not reduced unacceptably when a higher proportion of holes is incorporated, the limit of 55% for clay units and 60% for concrete aggregate units may be increased for masonry units that are used in the country having the national experience.

3. The combined thickness is the thickness of the webs and shells, measured horizontally across the unit at right angles to the face of the wall.

Figure 6.6

already air-dried, and multiplied by the factor δ as given in table 3.2 (of the code) to allow for the height and width of the units.'

Note: In the NAD it is indicated that the compressive strengths of units produced following the BS methods should be multiplied by 1,2 to convert them from wet tested values into equivalent air dry values for use with ENV 1996 -1-1.

♦ '*When the compressive strength of masonry units is quoted as the* **characteristic** *strength when tested in accordance with EN 772-1 this should be converted to the normalised compressive strength by changing the value of the strength, to the mean equivalent, using a conversion factor based on the coefficient of variation,..*'
and then proceed as for the case when the mean strength is quoted.

There are similar guidelines for situations in which the action effects result in compressive forces act parallel to the bed face and when the compressive strength of a special shaped unit is expected to have a predominant influence upon the masonry strength.

The constant 'K' is a boxed value depending on the Group number, type of mortar, type of unit and whether or not longitudinal joints exist in the masonry, e.g. as in a collar-jointed wall. The values of 'K' indicated Table 3 of the NAD are the same as those given in DD ENV 1996-1-1.

Guidance is also given for determining the characteristic compressive strength of unreinforced masonry with unfilled vertical joints and shell bedded* masonry.

***Note:** shell bedded masonry is defined as masonry in which the units are bedded on two general purpose mortar strips at the outside edges of the bed face of the units.

6.5.1.2   *Characteristic Shear Strength of Unreinforced Masonry* (f_{vk})

Where experimental evidence to establish the characteristic shear strength of unreinforced masonry is not available for a specific project or on a national database, the following relationships can be assumed:

Equation (3.4)

♦ f_{vk} = $f_{vko} + 0{,}4\sigma_d$ **When considering all joints**
       = $\boxed{0{,}065}\ f_b$ **satisfying the requirements of**
       ≥ f_{vko} **Clause 5.1.5** so as to be filled**
       = limiting value given in table 3.5   **with general purpose mortar.**

****Note:** Clause 5.1.5 states that:

   8 mm ≤ *thickness of bed and perpend joints* ≤ 15 mm
   for general purpose and lightweight mortars,   and

   1 mm ≤ *thickness of bed and perpend joints* ≤ 3 mm
   for thin layer mortars.

where:
f_{vk} is the characteristic shear strength of unreinforced masonry,
f_{vko} is the shear strength, under zero compressive stress, determined in accordance with EN 1052-3 or EN 1052-4 or, for general purpose mortars not containing admixtures or additions, obtained from table 3.5.

Table 3.5 of the code provides values of f_{vko} and limiting values of f_{vk} depending on the Group number (i.e. 1, 2a, 2b, or 3) and the designed compressive strength of the mortar. The boxed values in Table 3.5 are the same as those given in the NAD.

EN 1052-3 and EN 1052-4, which are currently in preparation are the codes relating to the shear strength of masonry and dpcs respectively.

σ_d is the design compressive stress perpendicular to the shear in the member at the level under consideration, using the appropriate load combination,

f_b is the normalised compressive strength of the masonry units, as described in 3.1.2.1 for the direction of application of the load on the test specimens being perpendicular to the bed face (see section 6.5.1.1 of this text).

Equation (3.5)

$$\begin{aligned} \blacklozenge \quad f_{vk} &= 0,5f_{vko} + 0,4\sigma_d \\ &= \boxed{0,045}\,f_b \\ &\geq f_{vko} \\ &= 0,7 \times \text{limiting value given in table 3.5} \end{aligned}$$

When using general purpose mortar and considering the perpend joints unfilled, but with the adjacent faces of the masonry units closely abutted together.

f_{vk}, f_{vko}, σ_d and f_b are as defined previously.

Equation (3.6)

$$\begin{aligned} \blacklozenge \quad f_{vk} &= \frac{g}{t}f_{vko} + 0,4\sigma_d \\ &= \boxed{0,025*}\,f_b \\ &\geq f_{vko} \\ &= 0,7 \times \text{limiting value given in table 3.5} \end{aligned}$$

When using shell bedded masonry, made with Group 1 masonry units and bedded on two equal strips of general purpose mortar, each at least 30 mm in width, at the outside edges of the bed face of the unit.

*Note: There is an error in equation 3.6 of the code where the value of 0,025 is given as 0,05.

g   is the total width of the two mortar strips,
t   is the thickness of the wall.

In addition, in Clause 3.6.3(6) the code states:
'For thin layer mortars, used with autoclaved aerated concrete units, calcium silicate or concrete units, the value of f_{vk} obtained from equations (3.4), (3.5) and (3.6), and the limits applicable to those equations, may be assumed, using the values given in Table 3.5 for clay units of the same Group and M10 to M20 mortar.'

The boxed values given in the code equations and in Table 3.5 are same as given in the NAD.

6.5.1.3   Characteristic Flexural Strength of Unreinforced Masonry (F f_{xk1}/f_{xk2})

The characteristic flexural strength is expressed in a similar manner to masonry strengths, i.e. F f_{xk1}/f_{xk2} N/mm^2 , where:

f_{xk1} is the characteristic strength in the plane of failure parallel to the bed joints, and
f_{xk2} is the characteristic strength in the plane of failure perpendicular to the bed joints.

e.g. **F 0,7/2,0** represents masonry with the strength equal to 0,7 N/mm^2 in a plane of failure parallel to the bed joints and a strength equal to 2,0 N/mm^2 in a plane of failure perpendicular to the bed planes.

As indicated in the NAD where test data relating to the flexural strengths in accordance with EN 1052-2* is not available, the values given in Table 3 of BS 5628 : Part 1 : 1992 should be used, in which case:

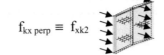

$$f_{kx\ par} \equiv f_{xk1} \qquad f_{kx\ perp} \equiv f_{xk2}$$

Figure 6.7

and the mortar designations given in Table 5 of the NAD, i.e. Types (i) to (iv) and M12 to M2 are used.

* **Note:** EN 1052-2 is the proposed code to determine the flexural strength of masonry which has still to be completed and published.

6.5.2 Mortar

Mortars are classified according to :

♦ their designed compressive strength expressed as the letter M followed by the magnitude of the compressive strength in N/mm^2 , e.g.

  M2,5 indicates mortar of compressive strength 2,5 N/mm^2
  M10 indicates mortar of compressive strength 10 N/mm^2

 or

♦ their prescription, e.g.

  1:1:5 indicates **(cement : lime : sand)** proportions by volume

In Table 5 of the NAD a list of strength classes to be associated to mortar designations (i) to (iv) as used in BS 5628 : Part 1 : 1992 are given as shown in Figure 6.8.

Extract from NAD Table 5:

Table 5. Strength classes to be ascribed to mortar designations to be used in ENV 1996–1–1				
Mortar Designation	(i)	(ii)	(iii)	(iv)
Cement : Lime : Sand	1 : 0 to ¼ : 3	1 : ½ : 4 to 4½	1 : 1 : (5 to 6)	1 : 2 : (8 to 9)
Masonry cement : Sand	–	1 : 2½ to 3½	1 : (4 to 5)	1 : (5½ to 6½)
Cement : Sand with plasticizer	–	1 : 3 to 4	1 : (5 to 6)	1 : (7 to 8)
Strength Class	M12	M6	M4	M2

Figure 6.8

The required strength of mortar is dependent on its intended use and as indicated in Clause 3.2, in each case should satisfy the following criteria:

♦ For *General Purpose* mortar:
    ≥ M1   in joints without reinforcement,
    ≥ M5   in joints containing reinforcement/prestressing steel,
    ≥ M2,5 in joints containing prefabricated bed joint reinforcement.

♦ For *Thin Layer* mortar:
    ≥ M5 and be designed in accordance with EN 998-2[*]
    Thin layer mortar is intended for use in masonry with bed joints with a nominal thickness of 1 mm to 3 mm.

♦ For *Lightweight mortar* :
    ≥ M5 and be designed in accordance with EN 998-2[*]
    Lightweight mortars should be made using perlite, pumice, expanded clay, expanded shale or expanded glass as the aggregate. It is acceptable to use other materials provided that tests are carried out to confirm their suitability.

* Note: EN 998-2 is the proposed code for mortar which has still to be completed and published.

The effects of the mortar strength on the characteristic compressive strength of masonry (f_k) are taken into account in the value of the constant 'K' referred to in section 6.5.1.1 of this text.

6.5.3   *Categories of Manufacturing and Construction Control*

In EC6 there are two categories of manufacturing control and three categories of construction control relating to masonry units and masonry construction. The categories for manufacturing control are given in Clause 3.1.1 and are:

'***Category I*** *may be assumed where the manufacturer agrees to supply consignments of masonry units to a specified compressive strength and has a quality control scheme, the results of which demonstrate that the mean compressive strength of a consignment, when sampled in accordance with the relevant part of EN 771 and tested in accordance with EN 772-1, has a probability of failing to reach the specified compressive strength not exceeding* 5%.'

'***Category II*** *should be assumed when the mean value of the compressive strength of the masonry units complies with the declaration in accordance with the relevant part of EN 771, but the additional requirements for Category I are not met.*'

Category I corresponds with the 'Special' category and Category II corresponds with the 'Normal' categories as defined in Table 4 of BS 5628 : Part 1 : 1992.

The categories of construction control are referred to as 'Categories of Execution Control' and are referred to in the 'informative' **Appendix G** of the code. This Appendix emphasises the need to employ *'qualified and experienced'* personnel for supervision and inspection of the work. In a footnote it is indicated that the definitions of categories of execution may be defined in the NAD when it is considered necessary. There are three categories referred to in the code 'A', 'B' and 'C', however, only 'A' and 'B' are used in the NAD; they are defined as following:

'Category "A" of execution control may be assumed when both of the following conditions are satisfied:
  - *regular inspection of the work is made by appropriately qualified persons independent of the constructor's site staff to verify that the work is being executed in accordance with the drawings and specification;*
    NOTE. *In the case of Design-and-Build contracts, the Designer may be considered as a person independent of the site organization for the purposes of inspection of the work.*
  - *preliminary compressive strength tests carried out on the mortar to be used indicate conformity to the strength requirements given in table 5 of this NAD and regular testing of the mortar used on site shows that conformity to the strength requirements given in table 5 is being maintained.'*

'Category "B" of execution control should be assumed when either or both of the conditions for category 'A' are not satisfied.
Category 'B' level of execution should not be used for reinforced or prestressed masonry, except in the following circumstances:
  - *deep beam, composite lintel construction and walls incorporating bed joint reinforcement to enhance lateral load resistance;*
  - *masonry containing prefabricated bed joint reinforcement used solely to control cracking.'*

Category 'C' execution control is not used in the NAD.

6.5.4   *Design of Masonry*

The design of structural members is similar to the methods adopted in BS 5628 : Part 1 : 1992, establishing expressions representing element resistance in terms of the effective thickness, slenderness ratio, capacity reduction factor, characteristic strength and partial safety factors for materials.

6.5.4.1   *Effective Thickness*   (Clause 4.4.5)

The effective thickness 't_{ef}' of single-leaf, collar-jointed (double-leaf), faced, shell-bedded, veneer and grouted cavity walls should be taken as the actual thickness.

In cavity wall construction in which both leaves are effectively connected by wall ties, the effective thickness should be calculated using:

$$t_{ef} = \sqrt[3]{t_1^3 + t_2^3}$$

where t_1 and t_2 are the thicknesses of the leaves.

In cases where the value of Young's Modulus (E) for the loaded leaf of a cavity wall is higher than the other leaf resulting in an overestimate of the effective thickness, the relative stiffness should be taken into account when calculating 't_{ef}'.

In addition when only one leaf of a cavity wall is loaded, and the wall ties are sufficiently flexible to ensure that the loaded leaf is not adversely affected by the unloaded leaf, the effective thickness can be calculated using this equation on the basis that the thickness of the unloaded leaf is not taken to be greater than the thickness of the loaded leaf.

This differs from the value determined using Figure 3 of BS 5628 : Part 1 : 1992, however, the difference is relatively small particularly in the case of a typical cavity wall built in the UK, e.g. consider a wall comprising two leaves each 102.5 mm thick:

Value of 't_{ef}' using BS 5628 : Part 1 : 1992   =   136.7 mm
Value of 't_{ef}' using EC6   =   129.1 mm

The NAD indicates that the effective thickness of walls stiffened by piers should be taken from Figure 3 and Table 5 of BS 5628 : Part 1 : 1992.

6.5.4.2 *Effective Height*   *(Clause 4.4.4)*

The effective height is determined by multiplying the clear storey height 'h' by a reduction factor 'ρ_n' where n equals 2, 3, or 4. The value of 'ρ_n' is dependent on the edge restraint or stiffening a wall. There are five cases considered:

Case I
'For walls restrained at the top and bottom by reinforced concrete floors or roofs spanning from both sides at the same level or by a reinforced concrete floor spanning from one side only having a bearing of at least 2/3 the thickness of the wall but not less than 85 mm:'

$$\rho_2 = \boxed{0,75} \quad \text{when } e \leq (0,25 \times t) \quad \text{and}$$
$$\rho_2 = \mathbf{1,0} \quad \text{when } e > (0,25 \times t)$$

where:
e   is the eccentricity at the top of the wall,
t   is the thickness of the wall.

Case II
'For walls restrained at the top and bottom by timber floors* or roofs spanning from both sides at the same level or by a timber floor spanning from one side having a bearing of at least 2/3 the thickness of the wall but not less than $\boxed{85}$ mm:'

$$\rho_2 = \boxed{1,0} \quad \text{when } e \leq (0,25 \times t) \quad \text{and}$$

where:
e   is the eccentricity at the top of the wall,
t   is the thickness of the wall.

*Note: In BS 5628 : Part 1 : 1992 an enhanced resistance to lateral movement can be assumed in the case of houses of not more than three storeys if a timber floor spans onto a wall from one side and has a bearing of not less than 90 mm; this is not recognised in EC6.

The boxed values are the same as those given in the NAD.

Case III
'When neither of the conditions given in Case I or Case II apply:'

$$\rho_2 = 1,0$$

Case IV
'or walls restrained at the top and bottom and stiffened on one vertical edge (with one free vertical edge):'
when h ≤ 3,5L,

$$\rho_3 = \frac{1}{1+\left[\dfrac{\rho_2 h}{3L}\right]^2}\rho_2$$

> 0,3   with ρ_2 from Cases I, II or III which ever is appropriate   or

when h > 3,5L,

$$\rho_3 = \frac{1,5L}{h}$$

where L is the distance of the free edge from the centre of the stiffening wall.

Case V
'For walls restrained at the top and bottom and stiffened on two vertical edges:'

when h ≤ L,

$$\rho_4 = \frac{1}{1+\left[\dfrac{\rho_2 h}{L}\right]^2}\rho_2$$

> 0,3   with ρ_2 from Cases I, II or III which ever is appropriate   or

when h > L,

$$\rho_4 = \frac{0,5L}{h}$$

where L is the distance between the centres of the stiffening walls.

Note: If L ≥ 30t, for walls stiffened on two vertical edges, or if L ≥ 15t, for walls stiffened on one vertical edge, where t is the thickness of the stiffened wall, such walls should be treated as walls restrained at the top and bottom only.

6.5.4.3   Slenderness Ratio   (Clause 4.4.6)

The slenderness ratio is defined as in BS 5628 : Part 1 : 1992 :

$$\text{Slenderness ratio} \;=\; \frac{h_{effective}}{t_{effective}} \;\leq\; \boxed{27}$$

The boxed value given in the NAD is the same as in EC6.

6.5.4.4   Out-of-Plane Eccentricity   (Clause 4.4.7)

The out-of-plane eccentricity can be estimated using the procedure given in 'Appendix C' of the code in which a simplified sub-frame of the structure is analysed to determine the value of the moment at a joint between walls and incoming beams as shown in Figure 6.9.

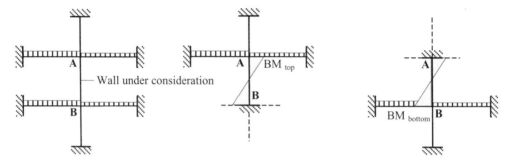

Remote ends from joint under consideration are assumed fixed unless they are known to take zero moment, i.e. pinned

Sub-frame used to evaluate the bending moment at the top of the wall

Sub-frame used to evaluate the bending moment at the bottom of the wall

Figure 6.9

The value of moment determined from the analysis is then used in combination with the axial load to determine the eccentricity. In multi-storey buildings the additional vertical load due to storeys above the level being considered should be added since this is not included in the sub-frame analysis.

The simplified frame model is not appropriate where timber floor joists are used; further guidance is given for use in these circumstances in the appendix. The NAD provides clarification on the use of Appendix C.

In addition to the eccentricity calculated from above, EC6 requires an *accidental eccentricity* 'e_a' to be included to allow for construction imperfections as indicated in Clause 4.4.7.2. The value of e_a should be taken as 'h_{ef} / $\boxed{450}$' The boxed value of 450 is also given in the NAD and applies to all categories of execution.

6.5.4.5   Slenderness Reduction Factor   (Clause 4.4.3)

The slenderness reduction factor ('β' in BS 5628 : Part 1 : 1992) allows for slenderness and/or load eccentricity. The critical section for design strength may be determined at the top or bottom of a wall, 'Φ_i' or at the mid-height of a wall, 'Φ_m'. The appropriate value of 'Φ' is calculated using equations '4.7' or '4.9' and Figures 4.1 and 4.2 in the code.

Glossary of Commonly Used Terms

Bat: A portion of brick manufactured or formed on site by cutting a whole brick across its length, e.g. a *snapheader* (see Chapter 1 Figure 1.3).

Bed face: The face of a structural unit which is normally laid on the mortar bed.

Bed joint: A mortar layer between the bed faces of masonry units.

Bond: An arrangement of structural units in an element (e.g. wall) designed to ensure that vertical, horizontal and transverse distribution of load occurs throughout the element (see Chapter 1 Figures 1.9 to 1.18).

Brick: A masonry unit, including joint material, which does not exceed 337.5 mm in length, 225 mm in width and 112.5 mm in height. In addition, to avoid confusion with tile work, the height should not be less than 38 mm.

Block: A masonry unit exceeding in length, width or height the dimensions specified for a brick. To avoid confusion with slabs and panels the height of a block should not exceed either its length or six times its width.

Brickwork/Blockwork: An assemblage of bricks or blocks bonded together to create a structural element.

Cavities: Holes which are closed at one end.

Cavity wall: Two parallel single-leaf walls, usually at least 50 mm apart, and effectively tied together with wall ties, the space between being left as a continuous cavity or filled with non-loadbearing material (usually thermal insulation).

Cellular bricks: Bricks having holes closed at one end which exceed 20% of the volume of the brick.

Chase: Channel formed in the face of masonry.

Closer: A portion of brick manufactured or formed on site by cutting a whole brick across its length, and used to maintain bond, e.g. *kingcloser* and *queencloser* (see Chapter 1 Figures 1.3, 1.13. 1.17).

Collar-jointed wall:	Two parallel single-leaf walls spaced at least 25 mm apart, with the space between them filled with mortar and so tied together as to result in composite action under load.
Common bricks:	Masonry unit suitable for general construction but with no particular surface finish or attractive appearance.
Coordinating size:	The size of a coordinating space allocated to a masonry unit, including allowances for joints and tolerances (see Chapter 1 Figure 1.5).
Corbel:	A unit cantilevered from the face of a wall to form a bearing.
Cornice:	A continuous projection from the facade of a building, part of a building or a wall.
Course:	A layer of masonry which includes a layer of mortar and masonry units.
Damp-proof course: **(dpc)**	A layer or layers, of material laid or inserted in a structure to prevent the passage of water.
Double-leaf wall:	See collar-jointed wall
Dowel:	A devise such as a flat strip or round bar, of uniform cross-section embedded in the mortar of some of the horizontal joints (beds) at the ends of a panel, and fixed rigidly to an adjacent structure to provide lateral restraint thus preventing in-plane horizontal movement of the panel.
Efflorescence:	The resulting white bloom left on the surface of a wall after soluble salts, which are present in the bricks/mortars, are washed out by excess water. These salts will subsequently re-dissolve in rain and be washed away in a relatively short period of time.
Engineering bricks:	Dense, semi-vitreous fired-clay bricks having minimum compressive strength and maximum absorption characteristics conforming to the requirements of BS 3921 : 1985.
Faced wall:	A wall in which the facing and backing are bonded such that they behave compositely under load.
Facing bricks:	Masonry units which are specially manufactured to provide an aesthetically attractive appearance.

Fair faced:	Work built with particular care, with respect to line and with even joints, where it is visible when finished.
Flashing:	A sheet of impervious material (e.g. lead, bituminous felt) applied to a structure and dressed to cover an intersection or joint where water would otherwise penetrate.
Frogged bricks:	Bricks having depressions formed in one or more bed faces, the volume of which does exceed 20% of the gross volume of the brick.
Grip hole:	A formed void in a masonry unit to enable it to be more readily grasped and lifted with one or both hands or by machine.
Grouted cavity wall:	Two parallel single-leaf walls, spaced at least 50 mm apart, effectively tied together with wall ties and with the intervening cavity filled with fine aggregate concrete (grout), which may be reinforced, so as to result in composite action under load.
Header:	A structural unit with its end showing on the face of the wall (see Chapter 1 Figure 1.9).
Hollow bricks:	Bricks having holes in excess of 25% and larger than perforated bricks.
Indenting:	The omission of structural units to form recesses into which future work can be bonded.
Jamb (Reveal):	The visible part of each side of a recess or opening in a wall.
Joint: ***Bed joint:***	The mortar layer upon which the structural units are set.
Cross joint:	A joint, other than a bed joint, normal to the face of the wall.
Wall joint:	A joint parallel to the face of a wall.
Jointing:	The filling and finishing of raked-out joints during construction (see Chapter 1 Figures 1.19 to 1.22).
Lime Bleeding:	The resulting staining left on the surface of a wall after soluble lime, which is produced during the hydration of the mortar, is deposited by the movement of rain water through freshly set and hardened mortar. These disfiguring stains will not weather off but will require removal with dilute acid to restore the appearance of the masonry.
Loadbearing masonry:	Masonry, which is suitable for supporting significant vertical/lateral, loads in addition to its own self-weight.

Longitudinal joint: A vertical mortar joint within the thickness of a wall, parallel to the face of the wall.

Movement joint: A joint specifically designed and provided to permit relative movement of a wall and its adjacent structure to occur without impairing the functional integrity of the structure as a whole.

Padstone: A strong block, usually concrete, bedded on a wall to distribute a concentrated load.

Panel: An area of masonry with defined boundaries which may or may not contain openings.

Partition: An internal wall intended for visual sub-division of space, i.e. non-loadbearing.

Perforated bricks: Bricks having holes in excess of 25% of the brick's volume, provided the holes are less than 20 mm wide or 500 mm^2 in area with up to three handholds within the 25% total.

Perpend joint: A mortar joint perpendicular to the bed joint and to the face of the wall (a vertical cross-joint).

Pigment: Inert mineral additives used to extend the colour range of mortar beyond that which can be achieved using various natural sands, cement and lime.

Pointing: The filling and finishing of raked-out joints after construction (see Chapter 1 Figures 1.19 to 1.22).

Quoin block: An external corner block (see Chapter 1 Figure 1.7).

Shell bedded wall: A wall in which the masonry units are bedded on two general purpose mortar strips at the outside edges of the bed face of the units (general purpose mortar as defined in EC6).

Shear wall: A wall to resist lateral forces in its plane.

Single-leaf wall: A wall of structural units laid to overlap (see bond) in one or more directions and set solidly in mortar.

Sleeper wall: A dwarf wall, usually honeycombed, to carry a plate supporting a floor.

Slip: A masonry unit either manufactured or cut, of the same height and length as a header or stretcher, and normally with a thickness of between 20 mm or 50 mm.

Solid bricks:	A brick which has no holes, cavities or depressions
Squint:	A special brick manufactured for an oblique quoin (i.e. on an external corner).
Stock bricks:	Brick originally hand-made in the south-east of England, so called from the timber 'stock' fixed to the bench that forms the '*frog*'. Sometimes used to describe bricks held in stock by brick-makers or merchants.
Strap:	A device for connecting masonry members to other adjacent components, such as floors and roofs.
Stretcher:	A structural unit with its length in the direction of the wall (see Chapter 1 Figure 1.9).
String course:	A distinctive course of brickwork in a wall, usually projecting from the wall and used as an architectural feature.
Structural units:	Bricks or blocks used in combination with mortar to construct masonry.
Stiffening wall:	A wall set perpendicular to another wall to give it support against lateral forces or to resist buckling and so to provide stability to the building.
Wall ties:	Metal strips used to connect the two separate leafs of cavity wall increasing the stiffness of each one.
Toothing:	Masonry units left projecting to bond with future work.
Veneered wall:	A wall having a face that is attached to the backing, but not so bonded as to result in composite action under load.
Weathering:	(a) the cover applied to, or the geometrical form of, a part of a structure to enable it to shed rainwater, (b) the effect of climatic and atmospheric conditions on the external surface of materials.
Wire-cut bricks:	Bricks shaped by extruding a column of clay through a die, the column being subsequently cut to the size of a brick.
Work size:	The size of a building component specified for its manufacture, to which its actual size should conform within specified permissible deviations (see Chapter 1 Figure 1.5).

Appendix A

Properties of Geometrical Figures

A =   Cross-sectional area

y or y_1 =   Distance to centre of gravity

Z_{xx} =   Elastic Section Modulus about the x-x axis

r_{xx} =   Radius of Gyration about the x-x axis

I_{xx}, I_{yy} =   Second Moment of Area about the x-x and y-y axes

Square:

$A = d^2$

$$I_{xx} = \frac{d^4}{12}$$

$$r_{xx} = \frac{d}{\sqrt{12}}$$

$y = d/2$

$$Z_{xx} = \frac{bd^3}{6}$$

Square:

$A = d^2$

$$I_{xx} = \frac{d^4}{3}$$

$$r_{xx} = \frac{d}{\sqrt{3}}$$

$y = d$

$$Z_{xx} = \frac{d^3}{3}$$

Square:

$A = d^2$

$$I_{xx} = \frac{d^4}{12}$$

$$r_{xx} = \frac{d}{\sqrt{12}}$$

$$y = \frac{d}{\sqrt{2}}$$

$$Z_{xx} = \frac{d^3}{6\sqrt{2}}$$

Rectangle :

$A = bd$

$$I_{xx} = \frac{bd^3}{12}$$

$$r_{xx} = \frac{d}{\sqrt{12}}$$

$y = d/2$

$$Z_{xx} = \frac{bd^2}{6}$$

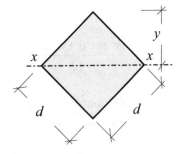

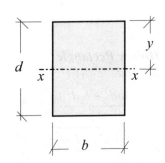

Rectangle:

$A = bd$

$y = d$

$I_{xx} = \dfrac{bd^3}{3}$

$Z_{xx} = \dfrac{bd^2}{3}$

$r_{xx} = \dfrac{d}{\sqrt{3}}$

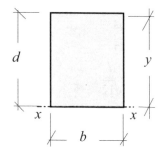

Rectangle:

$A = bd$

$y = \dfrac{bd}{\sqrt{b^2 + d^2}}$

$I_{xx} = \dfrac{b^3 d^3}{6(b^2 + d^2)}$

$Z_{xx} = \dfrac{b^2 d^2}{6\sqrt{b^2 + d^2}}$

$r_{xx} = \dfrac{bd}{\sqrt{6(b^2 + d^2)}}$

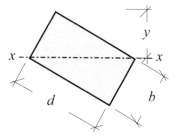

Rectangle:

$A = bd$

$y = \dfrac{b \sin\alpha + d \cos\alpha}{2}$

$I_{xx} = \dfrac{bd(b^2 \sin^2\alpha + d^2 \cos^2\alpha)}{12}$

$Z_{xx} = \dfrac{bd(b^2 \sin^2\alpha + d^2 \cos^2\alpha)}{6(b \sin\alpha + d \cos\alpha)}$

$r_{xx} = \sqrt{\dfrac{b^2 \sin^2\alpha + d^2 \cos^2\alpha}{12}}$

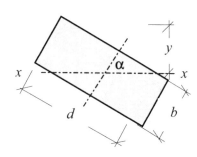

Hollow Rectangle

$A = (bd - b_1 d_1)$

$y = d/2$

$I_{xx} = \dfrac{(bd^3 - b_1 d_1^3)}{12}$

$Z_{xx} = \dfrac{(bd^3 - b_1 d_1^3)}{6d}$

$r_{xx} = \sqrt{\dfrac{bd^3 - b_1 d_1^3}{12A}}$

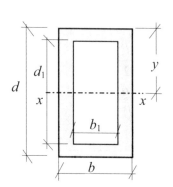

Trapezoid:

$$A = \frac{d(b+b_1)}{2}$$

$$y = \frac{d(2b+b_1)}{3(b+b_1)}$$

$$I_{xx} = \frac{d^3(b^2 + 4bb_1 + b_1^2)}{36(b+b_1)}$$

$$Z_{xx} = \frac{d^2(b^2 + 4bb_1 + b_1^2)}{12(2b+b_1)}$$

$$r_{xx} = \frac{d}{6(b+b_1)}\sqrt{2(b^2 + 4bb_1 + b_1^2)}$$

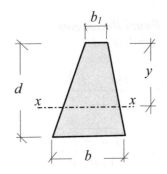

Circle:

$$A = \pi R^2$$

$$y = R = \frac{d}{2}$$

$$I_{xx} = \frac{\pi d^4}{64} = \frac{\pi R^4}{4}$$

$$Z_{xx} = \frac{\pi d^3}{32} = \frac{\pi R^3}{4}$$

$$r_{xx} = \frac{d}{2} = \frac{R}{4}$$

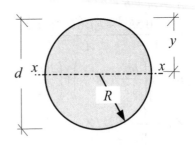

Hollow Circle:

$$A = \frac{\pi(d^2 - d_1^2)}{4}$$

$$y = R = \frac{d}{2}$$

$$I_{xx} = \frac{\pi(d^4 - d_1^4)}{64}$$

$$Z_{xx} = \frac{\pi(d^4 - d_1^4)}{32d}$$

$$r_{xx} = \frac{\sqrt{d^2 - d_1^2}}{4}$$

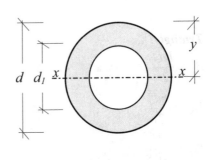

Semi-Circle:

$$A = \frac{\pi R^2}{2}$$

$$y = R\left(1 - \frac{4}{3\pi}\right)$$

$$I_{xx} = R^4\left(\frac{\pi}{8} - \frac{8}{9\pi}\right)$$

$$Z_{xx} = \frac{R^3(9\pi^2 - 64)}{24(3\pi - 4)}$$

$$r_{xx} = R\frac{\sqrt{9\pi^2 - 64}}{6\pi}$$

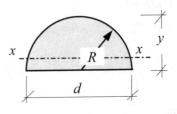

Equal Rectangles:

$$A = b(d - d_1) \qquad\qquad y = d/2$$

$$I_{xx} = \frac{b\left(d^3 - d_1^3\right)}{12} \qquad\qquad Z_{xx} = \frac{bd^3 - d_1^3}{6d}$$

$$r_{xx} = \sqrt{\frac{d^3 - d_1^3}{12(d - d_1)}}$$

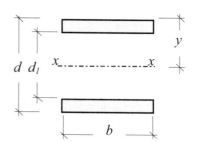

Unequal Rectangles:

$$A = bt + b_1 t_1$$

$$y = \frac{0.5bt^2 + b_1 t_1 \left(d - 0.5t_1\right)}{A}$$

$$I_{xx} = \left\{ \left(\frac{bt^3}{12} + btc^2 \right) + \left(\frac{b_1 t_1^3}{12} + b_1 t_1 c_1^2 \right) \right\}$$

$$Z_{xx} = \frac{1}{y} \qquad Z_{xx1} = \frac{1}{y_1}$$

$$r_{xx} = \sqrt{\frac{I_{xx}}{A}}$$

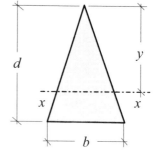

Triangle:

$$A = \frac{bd}{2} \qquad\qquad y = \frac{2d}{3}$$

$$I_{xx} = \frac{bd^3}{36} \qquad\qquad Z_{xx} = \frac{bd^2}{24}$$

$$r_{xx} = \frac{d}{\sqrt{18}}$$

Triangle:

$$A = \frac{bd}{2} \qquad\qquad y = d$$

$$I_{xx} = \frac{bd^3}{12} \qquad\qquad Z_{xx} = \frac{bd^2}{12}$$

$$r_{xx} = \frac{d}{\sqrt{6}}$$

Bibliography

1. **BDA Design Note 3**
   Brickwork Dimension Tables,
   Brick Development Association, Winkfield, Windsor, Berkshire, SL4 2DX

2. ***BS 187: Specification for calcium silicate (sandlime and flintlime) bricks***
   BSI, 1978

3. ***BS 648: Schedule of weights of building materials***
   BSI, 1964

4. ***BS 743: Specification for materials for damp proof courses***
   BSI, 1970

5. ***BS 1217: Specification for cast stone***
   BSI, 1997

6. ***BS 1243: Specification for metal ties for cavity wall construction***
   BSI, 1978

7. ***BS 3921: Specification for clay bricks and blocks***
   BSI, 1985

8. ***BS 4449: Specification for carbon steel bars for the reinforcement of concrete***
   BSI, 1997

9. ***BS 4729: Specification for dimensions of bricks of special shapes and sizes***
   BSI, 1990

10. ***BS 4887: Mortar plasticizers***
    BSI, Part 1:1986, Part 2:1987

11. ***BS 5224: Specification for masonry cement***
    BSI, 1995

12. ***BS 5262: Code of practice for external renderings***
    BSI, 1991

13. ***BS 5390: Code of practice for stone masonry***
    BSI, 1976

14. **BS 5628: Code of practice for use of masonry:**
     Part 1: Structural use of unreinforced masonry
     Part 2: Structural use of reinforced and prestressed masonry
     Part 3: Materials and components, design and workmanship
     BSI,   1992, 2000, 1985

15. **BS 6073: Precast concrete masonry units**
     Part 1 Specification for precast masonry units
     BSI, 1981

16. **BS 6399: Loading for buildings:**
     Part 1: Code of practice for dead and imposed loads
     Part 2: Code of practice for wind loads
     Part 3: Code of practice for imposed roof loads
     BSI,   1996, 1997,1988

17. **BS 6457: Specification for reconstructed stone masonry units**
     BSI, 1984

18. **BS 6649: Specification for clay and calcium silicate modular bricks**
     BSI, 1985

19. **BS 6744: Specification for austenitic stainless steel bars for the reinforcement of concrete**
     BSI, 1986

20. **Eurocode 1: Basis of design and actions on structures   DD ENV 1991 – 1 – 1**
     Part 1: Basis of design   (together with United Kingdom National Application Document)
     BSI, 1996

21. **Eurocode 6 : Design of masonry structures   DD ENV 1996 – 1 – 1**
     Part 1.1: General rules for buildings – Rules for reinforced and unreinforced masonry (together with United Kingdom National Application Document)
     BSI, 1996

22. **Extracts from British Standards for students of structural design**
     4th Edition
     BSI, 1998

23. **DD 34: Clay bricks with modular dimensions**
     BSI, 1974   (Withdrawn, replaced by BS 6649:1985)

24. ***DD 59: Calcium silicate bricks with modular dimensions***
BSI, 1978. (Withdrawn, replaced by BS 6649:1985)

25. ***DD 86: Damp proof courses:***
Part 1: Methods of test for flexural bond strength and short-term shear strength
Part 2: Method of test for creep deformation
BSI, 1983, 1984

26. ***DD 140: Wall ties:***
Part 1: Methods of test for mortar joint and timber frame connections
Part 2: Recommendations for design of wall ties
BSI, 1986, 1987

27. **Curtin W.G., Shaw G., Beck J.K. & Bray W.A.**
Brick Diaphragm Walls in Tall Single-Storey Buildings,
Brick Development Association, 1977.

28. **Curtin W.G., Shaw G., Beck J.K. & Bray W.A.**
Designing in Reinforced Brickwork,
Brick Development Association, 1983.

29. **Curtin W.G., Shaw G., Beck J.K. & Bray W.A.**
Loadbearing Brickwork Crosswall Construction,
Brick Development Association, 1983.

30. **Curtin W.G., Shaw G., Beck J.K. & Bray W.A.**
Structural Masonry Designers' Manual,
Second Edition, BSP Professional Books, 1987.

31. **Curtin W.G., Shaw G., Beck J.K. & Parkinson G.I.**
Structural Masonry Detailing,
Granad Publishing Ltd.,1984.

32. **Curtin W.G., Shaw G., Beck J.K. & Howard J.**
Design of Post-Tensioned Brickwork,
Brick Development Association, 1989.

33. **Freudenthal, A.M.**
The Safety of Structures
Proceedings of the American Society of Civil Engineers, October 1945.

34. **Hammett Michael**
A Basic Guide To Brickwork Mortars,
Brick Development Association, 1988.

35. **Hammett Michael**
Resisting Rain Penetration With Facing Brickwork,
Brick Development Association, 1997.

36. **Hammett M.**
Bricks - Notes on their properties,
Brick Development Association, 1999.

37. **Haseltine B.A. & Moore J.F.A.**
Handbook to BS 5628: Structural Use of Masonry: Part 1: Unreinforced Masonry,
Brick Development Association Design Guide 10, 1981.

38. **Haseltine B.A. & Tutt J.N.**
Handbook to BS 5628: Part 2: Section 1: Background and Materials,
Brick Development Association, 1991.

39. **Haseltine B.A. & Tutt J.N.**
Handbook to BS 5628: Part 2: Section 2: Reinforced Masonry Design,
Brick Development Association, 1992.

40. **Haseltine B.A. & Tutt J.N.**
Brickwork Retaining Walls,
Brick Development Association, 1981.

41. **Haseltine B.A. & Tutt J.N.**
External Walls: Design for Wind Loads,
Brick Development Association, 1984.

42. **Hendry A.W.**
The Calculation of Eccentricities in Load bearing Walls,
Brick Development Association.

43. **Hendry A.W.**
Structural Masonry,
Second Edition, MacMillan Press Ltd, 1998.

44. **Kaminetzky Dov**
Design and Construction Failures: Lessons from Forensic Investigations
McGraw-Hill, 1991.

45. **Korff J.O.A.**
Design of Freestanding Walls,
Brick Development Association, 1984.

46. **Morton J.**
Accidental Damage Robustness & Stability.
Brick Development Association, 1985.

47. **Parkinson G., Shaw G., Beck J.K. & Knowles D.**
Appraisal & Repair of Masonry,
Thomas Telford Ltd,, 1996

48. **Roark R.J.**
Formulas for Stress and Strain, 6ᵗʰ Edition,
McGraw-Hill, New York, 1989

49. **Timoshenko**
Strength of Materials: Part 2,
Van Nostrand Rheinhold, New York

50. **Johansen K.W.**
Yield Line Formulae for Slabs,
Cement and Concrete Association

51. ***Appraisal of Existing Structures***
Institution of Structural Engineers, 1996

52. ***Extracts from British Standards for students of structural design***
4ᵗʰ Edition,
BSI, 1998

53. ***Stability of Buildings***
Institution of Structural Engineers, 1988

54. ***The Collapse of a Precast Concrete Building under Construction***
Technical statement by the Building Research Station, London
HMSO, 1963.

55. ***Report of the Inquiry into the Collapse of Flats at Ronan Point, Canning Town, London***
HMSO, 1968.

56. ***Brick Cladding to Steel Framed Buildings***
Brick Development Association and British Steel Corporation, 1986.

57. ***Surveys and Inspections of Buildings and Similar Structures***
Institution of Structural Engineers, 1991

INDEX